INNOVATION
WAS NOT ENOUGH
———————————— A History of ————————————
the Midwestern Universities Research Association (MURA)

INNOVATION WAS NOT ENOUGH

— A History of —

the Midwestern Universities Research Association (MURA)

LAWRENCE JONES
University of Michigan, USA

FREDERICK MILLS
Fermi National Accelerator Laboratory, USA

ANDREW SESSLER
Lawrence Berkeley National Laboratory, USA

KEITH SYMON
University of Wisconsin-Madison, USA

DONALD YOUNG
Fermi National Accelerator Laboratory, USA

World Scientific

NEW JERSEY · LONDON · SINGAPORE · BEIJING · SHANGHAI · HONG KONG · TAIPEI · CHENNAI

Published by

World Scientific Publishing Co. Pte. Ltd.

5 Toh Tuck Link, Singapore 596224

USA office: 27 Warren Street, Suite 401-402, Hackensack, NJ 07601

UK office: 57 Shelton Street, Covent Garden, London WC2H 9HE

Library of Congress Cataloging-in-Publication Data
Innovation was not enough : a history of the Midwestern Universities Research
Association (MURA) / Lawrence Jones ... [et al.].
    p. cm.
  Includes bibliographical references and index.
  ISBN-13 978-981-283-283-2
  ISBN-10 981-283-283-1
1. Particle accelerators--Research--Middle West--History. 2. Midwestern
Universities Research Association. I. Jones, Lawrence W. (Lawrence William) 1925–
QC787.P3I516 2009
539.7'3072078--dc22

                             2008052809

British Library Cataloguing-in-Publication Data
A catalogue record for this book is available from the British Library.

Typeset by Stallion Press
Email: enquiries@stallionpress.com

Printed in Singapore.

CONTENTS

PREFACE

This book is a history of the Midwestern Universities Research Association (MURA). We much enjoyed writing it, for we all partook, to some degree, in MURA and were anxious to have the history of MURA properly recorded.

We would like to dedicate our book to the scientists, engineers, technicians, computer experts, secretaries, managers, administrators, university professors, university officials, and funding agency program managers, all of whom, collectively, made MURA what it was.

We wish to acknowledge the help of many people in the preparation of this book. They include, in no special order, Rebecca Kinraide, Esther Olson, Carl Baumann, Elizabeth Paris, Adrienne Kolb, Roy Kaltschmidt, Walter Trzciak, Michael Henelly, D. D. Reeder, the FNAL Photo Archives, the CERN Photo Archives, Jens Zorn, U. Minnesota Archives, David Z. Robinson of the Carnegie Commission in NY, Susan Mc Garvey, and the Purdue University Department of Physics Archives. We especially want to thank Joe Chew, who contributed greatly to the book by working through the proofs and making many corrections and many changes that vastly improved the presentation.

We have written not only for particle accelerator specialists who will be interested in the story of the MURA program, but for all those with an interest in the subject or in the history of science. Thus we have tried to keep the text, as much as possible, in a form which will be intelligible even to those with no prior knowledge of accelerator science.

This book could not have been written had we not so much enjoyed and wanted to augment and balance and then publish Frank Cole's work, "Oh Camelot!". We have drawn upon many books, reports, minutes, and conference proceedings, as well as upon personal recollections.

Keith R. Symon, Donald E. Young, Andrew M. Sessler,
Frederick E. Mills, and Lawrence W. Jones

AUTHOR'S BIOGRAPHIES

Keith R. Symon received his Ph.D. from Harvard University in 1948 and joined the faculty at Wayne University in Detroit. While at Wayne he wrote a popular text, *Mechanics*, and then joined the group that later became MURA, to work on particle accelerators. In 1955, he joined the faculty at the University of Wisconsin–Madison, where he became emeritus in 1990. He was on the staff of MURA from 1956 to 1967, and its Technical Director from 1957 to 1960. He was Acting Director of the Madison Academic Computing Center in 1982–1983, and Acting Director of the UW-Madison Synchrotron Radiation Center from 1983 to 1985. While with MURA, he developed the smooth approximation for the solution of equations with coefficients periodic in time, and invented FFAG accelerators. With A. M. Sessler, he formulated the theory of rf acceleration in fixed field accelerators. He was awarded the Particle Accelerator and Technology Award of the IEEE Nuclear and Plasma Science Society in 2003, and the American Physical Society Robert R. Wilson Prize in 2005.

Donald E. Young served as a Communication Officer in the infantry during World War II. Subsequently he did graduate studies at the University of Minnesota while working at the research laboratory of General Mills. After earning his doctorate in 1960, he joined the MURA laboratory with the mission of developing high-energy proton linear accelerators, extending his experience in the construction of the linac at the University of Minnesota. He spent the next six years in the development of linacs and, with the demise of MURA, chose to join the National Accelerator Laboratory (now Fermilab). As one of the first employees at Fermilab he was put in charge of building the 200 MeV linac injector. He retired from Fermilab in 1990 with a Scientist Emeritus appointment. He has been active since retirement as a consultant and in defense-related accelerator design. He is currently the President of the Particle Accelerator Corporation. He is also a Fellow of the American Physical Society.

Andrew M. Sessler received an A.B. from Harvard University and an M.A. and Ph.D. from Columbia University. After spending seven years on the faculty of Ohio State

Keith R. Symon, Donald E. Young, Andrew M. Sessler, Frederick E. Mills, and Lawrence W. Jones (from left to right).

University (during which time he interacted with MURA), he went to Lawrence Berkeley National laboratory (LBNL) in 1961 and has remained there ever since. He has published over 400 scientific papers, for which he has received a number of awards, including the Lawrence Award and the Wilson Prize. He is a former director of LBNL, a member of the National Academy of Sciences, and a former president of the American Physical Society. He has served on many national committees and has been active in arms control and human rights. For the latter he was the first winner of the Nicholson Award for Humanitarian Service. He is active in the Union of Concerned Scientists and in Amnesty International.

Frederick E. Mills received his baccalaureate and graduate education from the University of Illinois at Urbana-Champaign from 1946 to 1955. He was a Postdoctoral Fellow at Cornell University and then a Professor of Physics and Nuclear Engineering at the University of Wisconsin at Madison. He worked at Fermilab for 19 years, culminating in the Tevatron Collider, which is still at the forefront of particle physics. He also worked at MURA, where he was Director in 1965–1967; Brookhaven National Laboratory; Argonne National Laboratory, where he was a University of Chicago Argonne Fellow in 1990–1992; and Centre des Etudes Nucleaire de Saclay (France). He has served on many national committees. He has a number of patents to his name, including a synchrotron radiation system and a proton cancer therapy scheme. Following retirement from Fermilab and Argonne in 1993, he has been a consultant on muon accelerators and colliders, proton drivers, antiproton collectors, and medical accelerators. He is a Fellow of the American Physical Society. He has been a winter resident of Sun City, Vistoso, near Tucson, since 1993.

Lawrence W. Jones, following US Army service in World War II, received his B.S. and M.S. at Northwestern University and his Ph.D. at the University of California,

Berkeley. In 1952, he joined the University of Michigan's Physics Department, where he remained throughout his career, serving as Department Chair for a term. He was involved with the MURA group throughout the 1950s. At Michigan, he developed the luminescent chamber and has carried out experiments at the Berkeley Bevatron, the BNL Cosmotron and AGS, and the CERN PS. During the mid-1960s, he turned his attention to cosmic ray research, establishing a facility at Echo Lake, Colorado. He returned to accelerator experiments at Fermilab and later at LEP (CERN). More recently, he has been involved with detector developments for the CERN LHC and with the Armenian cosmic ray research program. A Fellow of the American Physical Society, he is author or coauthor of over 400 journal publications and over 200 papers in conference proceedings. He has spent sabbatical and research leaves in England, India, Australia, New Zealand, and Switzerland.

CHAPTER 1

INTRODUCTION

One day Mervyn Hine of CERN and I were sitting in the Fermilab coffee lounge and I grumbled, "The way people talk about MURA, you'd think it was Camelot." He answered very earnestly, "Oh, it was, Frank, it really was!"

— Francis Cole

This book presents a history of the Midwestern Universities Research Association (MURA). Why should one be interested in the history of MURA? After all, it lasted only from 1953 to 1967; and failed in its mission to obtain an accelerator for the Midwest. Nevertheless, there are many good reasons for reviewing its history.

First, 15 universities formed a confederation and jointly sponsored an activity that was larger than any one university could undertake. A similar gathering of Eastern universities had earlier created Brookhaven National Laboratory and had been highly successful. What motivated these universities and how was this joint activity accomplished? What was the interaction between MURA and the Atomic Energy Commission and its Argonne National Laboratory? Why did the Midwestern activity fail where the Eastern one succeeded? What was the lasting impact of the Midwestern joint activity? Some of this has been previously discussed [Greenberg, 1999; Greenbaum, 1971; Paris, 2003].

Second, the MURA group produced accelerator inventions, concepts, theoretical advances, and models that made it, arguably, the most prolific accelerator group that ever existed. The major contributions to accelerator science — to be described in detail in the book — are the concept of the fixed field alternating gradient as incorporated in the Radial Sector Model (Fig. 1.1) and the Spiral Sector Model (Fig. 1.2), and the realization of colliding beams as incorporated in the 50 MeV Two-Way Model (Fig. 1.3).

Third, in detail, what were the technical accomplishments? What has been the impact of these advances in the subsequent years? Why was the MURA group able to produce so much advanced research?

The scientific themes (such as radio frequency theory, instability theory, instability experimental studies, or particular models) will each be discussed completely

1

Fig. 1.1.   The Radial Sector Model at Michigan, 1956.

Fig. 1.2.   The Spiral Sector Model at Madison, 1958.

Fig. 1.3. The 50 MeV Two-Way Model at the Stoughton site, 1961.

as of their year of major activity. The book describes the innovations not in technical detail — those results have long since been published and incorporated into the tool kits of beam physicists — but in a manner which the authors hope will allow the reader to fully appreciate the concepts involved. This book describes the background of the scientific work — material that does not make it into scientific publications, but is vital if one is to appreciate how research is accomplished. Similarly, it is important to understand and describe the background against which MURA operated: the political and sociological aspects of those times. That includes developments in the Cold War, the policy of AEC, and technical developments to which MURA was unable to respond.

The Technical Director of MURA, during its formative years, was Donald W. Kerst. He was deeply involved in all aspects of MURA and the true leader of the group. A sidebar gives more information on this key figure.

The authors were all involved with MURA, spanning its full history, and have supplemented their memories with archival documents. At the end of this book are listed all of these historical items of which the authors are aware, and we describe how the interested reader can obtain these documents. There have been a number of articles on MURA's history. Some short pieces, by Lawrence Jones [Jones, 1991], Keith Symon [Symon, 1991], and Donald Kerst [Kerst, 1991], appear in the

Donald William Kerst (1911–1993)

Donald William Kerst was one of this country's most influential particle physicists among those who were educated just before World War II, when particle physics began. He made important contributions to the design of particle accelerators, to nuclear physics, to medical physics, and to plasma physics. Most particularly, he was the guiding light of MURA.

He was born in Galena, Illinois and was educated at the University of Wisconsin, where he received a B.A. in 1934 and a Ph.D. in 1937. His thesis research involved the development and application of an electrostatic generator for a seminal experiment on the scattering of protons by protons. After his degrees and a one-year interlude working on X-ray tubes and machines at the General Electric X-ray Corporation, Chicago, Kerst was challenged by high-energy electron and X-ray research, which required energies not yet available. He accepted an instructorship in 1938 at the University of Illinois, Champaign-Urbana, where he was encouraged by the prescient chairman of the Department of Physics, F. Wheeler Loomis, to develop his ideas for a new type of electron accelerator, which Kerst later named the "betatron."

Among the investigators who attempted to accelerate electrons by magnetic induction, none were successful until Kerst produced 2.3 MeV electrons in a betatron at the University of Illinois on July 15, 1940. That tabletop machine is now in the Smithsonian Museum in Washington, DC. All later accelerators, including the newest high-energy synchrotrons, have been influenced by this early work of Kerst.

Kerst later constructed a number of betatrons of successively higher energies, culminating in the 300 MeV betatron with its 400-ton magnet at Illinois. These machines were used by Kerst and his students to carry out much of the first research in the multi-MeV energy range. He also pioneered the use of MeV photon radiation in the treatment of cancer.

From 1943 to 1945, Kerst was the leader of the P-7 group at Los Alamos, which developed the first homogeneous fission reactor ("water boiler"). He also used the betatron to study the implosion method for igniting nuclear weapons. The official history of Los Alamos describes Kerst's work as follows: "The technical achievements are among the most impressive at Los Alamos."

While on leave from the University of Illinois from 1953 to 1957, Kerst served as Technical Director of MURA in Madison. His activities during this period are described in this book. His deep understanding of the physics of electric and magnetic fields and of mechanics, and his vigorous leadership, were responsible in large part for the many contributions to accelerator technology made by the MURA group during this period.

(Continued)

Donald William Kerst (*Continued*)

After 19 years at Illinois developing accelerators, Kerst accepted in 1957 an appointment of five years to work on plasma physics in the fusion program at the General Atomics Division of the General Dynamics Corporation in La Jolla, California. He brought to this field not only his deep physical insight into magnetic field structures, but also his under-standing, gained from his accelerator experience, of the importance of careful attention to detail in the design of magnetic structures so as to eliminate all possible sources of error and asymmetry in the magnetic fields. He introduced into plasma physics the advanced dynamical concepts developed at MURA. These are largely responsible for the success of the various toroidal machines that have been built under his direction, including a toroidal pinch device at General Atomics, and a number of multipole machines (of which he was coinventor with Tihiro Ohkawa).

In 1962, Kerst returned to the University of Wisconsin–Madison to establish a plasma pro-gram. The first multipole machines were the toroidal octupoles completed at the University of Wisconsin under his direction and the toroidal octupole started by him and Ohkawa at General Atomics and completed by the latter.

In addition to his scientific and technical contributions to accelerator, nuclear, medical, and plasma physics, his deep understanding, his know-how and enthusiasm have been a source of education and inspiration both to his students and to his colleagues. He was an enthusiastic and effective mentor who worked hard and expected his students to do like-wise, and they did. Thirty-three students completed their Ph.D. degrees in the betatron group at Illinois over a period of 30 years; 42 students completed their doctorates in the plasma group at the University of Wisconsin during the 17 years that Kerst led the group. Many of the leading scientists in accelerator, nuclear, medical, and plasma physics received their degrees under Kerst's direction.

Donald Kerst was a well-rounded person. He was a dedicated husband to his wife, Dorothy, and a loving father to his children. He was a sportsman who enjoyed skiing, deep-sea fish-ing, white-water canoeing, and ocean sailing. His low-key sense of humor often delighted his friends and colleagues.

(Based upon an obituary by Keith R. Symon and H. William Koch, *Physics Today*, Vol. 47, No. 58, Jan. 1994.)

Terwilliger Symposium. A short piece by Fred Mills [Mills, 2001], appears in the proceedings of the Cyclotrons, 2001 Conference. Francis Cole has written a rather extensive history [Cole, 1994], but that work suffers both from being very personal and from never having been published. We draw heavily upon these contributions, and also upon a rather extensive set of notes, reports, minutes, and extensive per-sonal recollections.

The authors have enjoyed looking back at those Camelot years, and trust that the reader will share the pleasure of looking, in some detail, at the brief span of time that was MURA.

CHAPTER 2

HISTORICAL BACKGROUND

2.1. THE EARLY HISTORY OF ACCELERATORS

In order to understand the contributions that MURA made to accelerator physics, it is necessary to think back to those times and recall what was known prior to MURA. Accelerator physics grew out of scientists' considerable success in the 1920s with generating ever-higher voltages, which could be used for electrostatic acceleration of particles. Considerable success was accomplished with this simple motivation. At the same time Ernest Rutherford, having obtained nuclear reactions using natural isotopes, called for the development of machines capable of "natural radioactivity," i.e., nuclear reactions. This motivation prompted considerable effort in many places. The first success was with electrostatic machines. Cockcroft and Walton in Cambridge, in 1932, were the very first to make nuclear reactions artificially. The generator they developed was widely used until the invention of, and development of, radio frequency quadrupoles (RFQs) long after the MURA days.

Shortly after the success of Cockcroft and Walton, Robert Van de Graaff, at the Department of Terrestrial Magnetism of the Carnegie Institute in Washington, and later at MIT, developed another type of electrostatic generator, which employed a mechanical belt to do the work necessary to build up the charge on an upper terminal. He reached 1.5 million electron-volts (MeV) in 1931 and, unlike Cockcroft and Walton, he attempted no nuclear reactions. At this time Merle Tuve, Gregory Breit, and Odd Dahl were also developing electrostatic machines at the Carnegie Institute. Electrostatic machines kept being improved and were used for many nuclear physics studies in the 1930s. Particularly, Raymond G. Herb of the University of Wisconsin introduced the pressurized electrostatic accelerators that quickly yielded a voltage of 4 MV and, ultimately, 25 MV.

Simultaneous with this work was the development of the cyclotron by Ernest Lawrence, at Berkeley. He had developed the concept in 1928 and 1929 and started construction in 1930, and the first machine was completed by Stanley Livingston in 1932. The cyclotron employed a magnetic field to keep the particles in a circular orbit while a radio frequency field increased their energy on each turn. (This was in marked contrast with the electrostatic machines described above. Electrostatic machines must impart the full energy change to the particles in one pass across an extremely high voltage difference. In the cyclotron, the particles' energy was increased incrementally with repeated applications of a small voltage, rather than all in one pass.) The idea was stimulated by the work of Rolf Wideroe. Soon Lawrence and many others were making cyclotrons, and these were significant tools in nuclear physics research in the 1930s. They appeared limited to nonrelativistic energies, as had been shown by theorists Hans Bethe and Morris Rose, although L. H. Thomas had shown in a rather dense paper (carefully studied by Lawrence's student Leonard Schiff) how to circumvent the limit. Nevertheless, Lawrence started to build the 184-inch cyclotron, but World War II intervened.

A third type of accelerator was under development in those early years, namely the linear accelerator. Ising had proposed, in 1922, the concept of using powerful radio waves to excite hollow metal cylinders, which alternated with "drift tubes" that shielded the particles from the opposite excitation until the next upswing of the radio-frequency energy. In this way a small voltage could be cascaded and a particle raised to a high energy without having (as in the electrostatic machines) a high voltage. Wideroe was the first person to take this idea and try to realize it in hardware. He did that for his thesis and the work was published in 1928, although it must be noted that the device did not produce sufficient energy to cause nuclear reactions. Sloan, in Lawrence's laboratory, made a hadron linac in 1931. This field was rather quiet until after WWII when, using the microwave developments of the war, Luis Alvarez made a 32 MeV proton linac. Soon many machines, modeled on Alvarez's work, were constructed throughout the world. Electron linacs were developed by Robert Hansen *et al.* at Stanford after World War II. Heavy ion accelerators were being developed just at the time of MURA; the first, the HILAC at Berkeley, was started in 1957 and completed in 1965 (and a similar machine was developed at Yale as the other half of a joint project).

Yet another way to accelerate electrons, besides sheer high voltage (as in electrostatic accelerators) and setting up electromagnetic fields with radio-frequency power (as in cyclotrons and linacs), is through induction. The beam of charged particles may be likened to the secondary of a transformer. The first success, a circular accelerator called the betatron (because it used "beta particles"), was produced by Donald Kerst in 1940 at Illinois.

At the end of WWII, Edwin M. McMillan and Vladimir Veksler both arrived at the concept of phase focusing. This concept allowed the accelerating rf voltage to be frequency-modulated. This led to the synchrocyclotron, which, like the cyclotron, has a constant magnetic field, but in which the particles are accelerated in bunches to a higher energy than in a cyclotron. More importantly, phase focusing allowed the development of synchrotrons, in which the magnetic field is increased as the bunch is accelerated. The concept was rapidly tested, at Berkeley, on a 320 MeV electron synchrotron, and then incorporated into the 184-inch proton cyclotron, converting it to a synchrocyclotron. Other proton machines soon followed: the 1 BeV (billion electron-volt — now called GeV, for giga-electron-volt) machine at Birmingham (1953), the 3.3 BeV Cosmotron (1953) at Brookhaven, the 6.2-BeV Bevatron (1954) at Berkeley and the 10-BeV Synchro-Phasotron at Dubna, Soviet Union (1957).

So, at the time MURA began, there were electrostatic machines, cyclotrons, linacs, betatrons, and synchrotrons. These machines were not only being used for nuclear physics (and later for particle physics), but also for a number of applications. For example, the Sloan linac (really just a pulsed transformer) was used in hospitals for therapy in the 1930s. Also, linacs and electrostatic machines were being employed for the same purpose. Even cyclotrons were used to explore hadron therapy — the use of protons or heavy ions to precisely kill tumors. The betatron was employed for medical and industrial purposes, and even during WWII, at Los Alamos, to produce penetrating X-rays.

It is important to understand the state of understanding and the technological ability at the time of MURA. Accelerators were certainly being built and operated successfully, but theoretical understanding was very elementary prior to WWII. Transverse focusing (both magnetic and electric) had been discovered, by serendipity, in cyclotrons. Accelerator builders judiciously put in shims on the falling magnetic field at the cyclotron edge and they observed that the accelerating "Dee" could produce vertical focusing. In order to make the betatron operate, Kerst and Serber had to analyze transverse focusing and the result was almost the first set of equations to enter the accelerator builder's tool kit. As a result of this important work, oscillations about the equilibrium orbit in all accelerators are now generally referred to as betatron oscillations. In order to develop the phase focusing concept, McMillan developed the equations for particles subjected to a slow frequency variation. Pretty much that was the theoretical understanding — even all that anyone thought they needed — prior to 1950: a few differential equations.

Since the MURA developments depend very much on the focusing of particles in magnetic fields, let us discuss focusing here in a little more detail. In circular particle accelerators, cyclotrons, betatrons, synchrotrons, etc. each particle moves many times around a nearly circular orbit. For each particle at a certain energy, there is a circular orbit — the equilibrium orbit in the midplane of the magnet — on which the particle should move. However, there are always small errors, so a particle will not remain exactly on the equilibrium circle. If there were no focusing, the particle would soon drift far from the equilibrium orbit and be lost. One therefore must provide a focusing force which pushes the particle back toward the equilibrium orbit, in both the horizontal and the vertical direction, thus keeping it near the equilibrium orbit.

In the early cyclotrons, focusing came about more or less accidentally. The magnetic field in a cyclotron is produced by a magnet having two circular pole faces — a north pole and a south pole — spaced a small distance apart. The magnetic field is nearly uniform and directed vertically, but decreases slightly in magnitude as radius increases, particularly near the edges of the magnet poles. This magnetic field gradient can be shown to produce a small vertical focusing, which keeps particles near the median plane. Unfortunately the same field gradient produces radial defocusing, which would tend to drive particles radially away from the equilibrium circle. This is a general property of magnetic fields; a magnetic gradient that focuses in one dimension is defocusing in the perpendicular dimension. It happens that there is another small radially focusing force due to the centrifugal force. If the magnetic field gradient is small enough so that the radial defocusing is less than the centrifugal focusing, one can have weak focusing both horizontally and vertically. Until the 1950s, all circular particle accelerators were weak focusing.

2.2. ACCELERATOR PHYSICS IN THE MIDWEST

Following WWII, with the invention of the betatron and the synchrotron, and the many other electronic and technical advances made during the war years, physicists at the Midwestern universities, as those at universities and laboratories elsewhere, built state-of-the-art particle accelerators on their campuses. Thus, by 1952, there was the Illinois 300 MeV betatron, the University of Chicago 450 MeV synchrocyclotron and 50 MeV betatron, the 300 MeV electron synchrotron at Purdue, the 68 MeV proton linear accelerator at Minnesota, the 70 MeV electron synchrotron at Iowa State University, and the 70–140 MeV electron synchrotron with straight sections at Michigan. There were also some prewar lower-energy accelerators, such as an advanced Van de Graaff accelerator at Wisconsin and a 40-inch cyclotron at Michigan. The work most relevant to MURA was that at Michigan.

In the late 1940s, after learning of the invention of the synchrotron concept and phase stability, H. Richard (Dick) Crane at the University of Michigan, in designing an electron synchrotron, developed the concept of straight sections, a departure from the azimuthal isotropy of all cyclotrons, betatrons, and electron synchrotrons up to that time. He included four straight sections in the electron synchrotron he designed, to provide easier access for the rf acceleration system, injection, reaction targets, etc. David Dennison, also at Michigan, with his student Ted Berlin made the calculations of the betatron oscillations, using the Hill equations (with matrices representing the displacements and slopes of the particle trajectories through the magnets and across the straight sections). Dennison later noted that, had he tried negative-focusing magnets and included them in his Hill equation calculations, he might have invented alternating gradient focusing.

Crane built his electron synchrotron at the University of Michigan; it had four straight sections and a 1 m radius of curvature in the magnet quadrants, so that it had a potential maximum energy of about 300 MeV. With a room-filling Cockroft–Walton 500 keV injector power supply, it operated successfully at this potential in the early 1950s; however, for most of its useful lifetime, it operated at about 70 MeV. While the synchrotron was still under construction, Crane and his students used the Cockroft–Walton electrons to make measurements of the g factor of the free electron — a major pioneering research accomplishment. Having just completed their degrees on the Berkeley 300 MeV synchrotron, Lawrence Jones and Kent Terwilliger were hired onto the Michigan faculty (as instructors) by Crane to be part of his synchrotron research team.

Accelerator builders were good at constructing magnets, but they had no computational ability, so only the simplest geometry could be considered and saturation effects in iron were treated very empirically. Radio frequency manipulation had been well developed in WWII. The technology of high vacuum, used in the electron tube industry, was not widely known. In fact Don Kerst went to General Electric to learn that technology, which he employed in the betatrons he built. Accelerator builders knew how to make electron and ion sources, but because there was no computational ability, understanding was limited to low-intensity sources. Injection of beams into accelerators was "black magic." For example, Kerst did not understand how a beam was injected into the betatron. In short, the technology of the day was severely constrained by the limited ability to perform extensive numerical calculations.

2.3. THE COMING OF STRONG FOCUSING

While building the Cosmotron at Brookhaven National Laboratory (BNL), Stan Livingston asked what would be the effect of having the C-shaped magnets alternating, i.e., some facing outward and some facing inward. Ernest Courant, analyzing this case, soon realized that a sequence of equally strong alternately focusing and defocusing magnets would have a net focusing effect. By using magnets with intense gradients, transverse focusing of particles both radially and vertically can be achieved to a much greater degree than in any weak focusing machine. Hartland Snyder immediately understood this effect, and could see its generalization, and he and Ernest quickly developed elegant formalisms for studying alternating gradient (AG) focusing, or "strong focusing," which they published in 1952. Since then, nearly all circular accelerators have been strong-focusing. The departure from azimuthal symmetry (introduction of straight sections) had been achieved earlier by Crane, but without appreciation of the generality of strong focusing.

Independently, Nicholas Christofilos had discovered strong focusing, but as he was then still an engineer building elevators in Athens, his noteworthy career in physics still largely in the future, his contribution was ignored and, unfortunately, it had no impact on developments.

That lack of impact was certainly not true of the work of Courant, Livingston, and Snyder. The very first use was by Robert Wilson, on a synchrotron at Cornell. Strong focusing was incorporated into the Alternating Gradient Synchrotron (AGS). And a delegation from CERN, hearing all about this development, went back to Switzerland and convinced their colleagues to change the design of the accelerator they would construct, the Proton Synchrotron (PS), to incorporate strong focusing. John Blewett, and others, realized that the concept would greatly improve linacs, and the idea of spatially separated quadrupole focusing magnets, rather than grids, was quickly incorporated into linacs, the first being the Heavy Ion Linear Accelerator (HILAC) at Berkeley.

The concept had a large impact upon MURA. One could say that the strong focusing concept introduced a whole new stage. All sorts of possibilities became available. Thus there would be developed straight sections, low beta sections, final focus sections, isochronous rings, lattices particularly suitable for synchrotron rings, and, also, the Fixed-Field Alternating Gradient (FFAG) concept. For further discussion of the development of particle accelerators, see the book *Engines of Discovery* [Sessler, 2007].

2.4. THE DESIRE FOR A NEW ACCELERATOR LABORATORY IN THE MIDWEST

The establishment of Brookhaven National Laboratory with its managing corporation, AUI (Associated Universities Incorporated; essentially the Ivy League schools), and the construction of the Brookhaven Cosmotron in the East (soon to be supplanted by the Alternating Gradient Synchrotron), plus the postwar evolution in Berkeley of the Radiation Laboratory, with its synchrocyclotron, electron synchrotron, proton linear accelerator, and (under construction) Bevatron in the West, stimulated Midwestern university senior scientists interested in these areas of physics to yearn for a central, federally funded research center in their area where a major, high-energy accelerator could be built.

It was during this period, the late 1940s and early 1950s, that high-energy elementary particle physics emerged as an identifiable field of physics, evolving from a merger of elements of nuclear physics and cosmic ray physics. The prewar particle accelerators (electrostatic accelerators — the betatron and cyclotron) had been developed for nuclear physics research. However, the discovery of muons, pions, and strange particles (kaons and hyperons) in cosmic rays at higher energies stimulated the community to build accelerators with energies above 100 MeV capable of producing these new particles in the laboratory. But the cost, size, and infrastructure required for accelerators with energies above a GeV were larger than a single university could easily accommodate, and it was logical that such a facility should be built at a regional laboratory. Of course, there was already a regional laboratory in the Midwest, Argonne National Laboratory, operated by the University of Chicago under contract with the AEC. However, its program, in those days (early 1950s), was devoted entirely to nuclear engineering. Access was limited by security measures. And the other Midwestern universities had no voice in its management or programs.

At the December 15, 1952, Cosmotron dedication event at Brookhaven, Samuel K. Allison (University of Chicago) and P. Gerald Kruger (University of Illinois) suggested that a meeting of scientists in the Midwest should be convened to consider ways of providing a high-energy facility for that part of the country, and to educate this community in the new alternating-gradient focusing concept and related technologies for high-energy accelerator design.

CHAPTER 3

THE EARLY MURA YEARS, 1953–1956

3.1. THE BEGINNINGS OF MURA

The meeting suggested by Allison and Kruger was held at the University of Chicago on April 17–19, 1953; this meeting might be considered the informal beginning of what became MURA. Speakers at the meeting included Ernest Courant and John Blewett (Brookhaven), and Robert R. Wilson (Cornell), who discussed alternating gradient theory as well as pragmatic issues of accelerator magnets and construction, etc. Wilson even brought and showed off a strong-focusing magnet lamination for the 1 GeV alternating gradient electron synchrotron he had already begun to construct. A large, representative group of Midwestern physicists were present, including Enrico Fermi. From the University of Michigan, H. Richard Crane brought with him Lawrence W. Jones and Kent M. Terwilliger, both first-year instructors. Jones recalls showing Fermi a pair of strong-focusing Lucite optical lenses he had made (planoconvex and planoconcave cylindrical lenses cemented together with their axes perpendicular), which Fermi found amusing.

The senior representatives of the Midwestern universities present formed an Organizing Committee, took steps to provide a continuation of activity, and asked Donald W. Kerst (of Illinois) to be the technical director and to explore the possibility that a group of Midwestern physicists could spend some time at Brookhaven that summer to educate themselves in these new concepts. The Brookhaven management enthusiastically supported this summer study. During these first years, the group called itself the Midwest Accelerator Conference (MAC).

Consequently, this MAC group spent July 7–21, 1953 at Brookhaven. They were lodged together in the Wyandotte Hotel at Bellport, and had a very productive two weeks. The participating physicists were Lawrence Johnston (University of Minnesota), Daniel Zaffarano and L. Jackson Laslett (Iowa State University), Jones and Terwilliger (University of Michigan), Francis T. Cole (University of Iowa), S. Courtney Wright (University of Chicago), Norman Francis (Indiana University), John Powell (University of Wisconsin), and Kerst. They had extensive lectures by Courant, Hartland Snyder, M. Stanley Livingston, Milton

White, Robert Serber, and Kenneth Green. Besides lectures, they were given tours of the Cosmotron, with detailed examination and discussion of the engineering aspects. The group continued its discussions at the Wyandotte Hotel in the evenings, so that this was a period of total immersion by those physicists. Much of this group reconvened at the University of Wisconsin from August 7 to September 5, joined by others including Fritz Rohrlich (University of Iowa), John Williams (University of Minnesota), Ragnar Rollefson (University of Wisconsin), and Crane. Among those who also joined the group in Madison was Keith R. Symon (Wayne University). Courant (Brookhaven) plus Robert Hofstadter and Wolfgang Panofsky (Stanford) also joined the group for a period (Fig. 3.1). At these discussions, much effort was directed toward orbit problems, e.g., nonlinearities, resonances, and phase oscillations. Some of the more pragmatic discussions included the conceptual design of a 10 GeV electron alternating gradient accelerator by Laslett and Jones.

During this period — and subsequently — the group participated in the exchange of internal reports and studies with other labs, especially Brookhaven and CERN (Geneva, Switzerland), labs which were especially active in the study of alternating gradient accelerators. It was decided to maintain the cohesion and

Fig. 3.1. The 1953 MAC Summer Study group at Madison. Standing, L-R: Norman Francis, Lawrence Jones, Kent Terwilliger, Courtney Wright, Ragnar Rollefson, Fritz Rohrlich, Frank Cole, and Donald Kerst. Seated, L-R: L. Jackson Laslett, Kim Wright, Wolfgang K. H. Panofsky, Robert Hofstader, Lawrence Johnston (in chair), Ernest Courant, John Williams, and Sarah Courant.

activity of the MAC group by holding weekend meetings every month or two at different Midwestern universities, hence there was a two-day meeting at the University of Illinois in October 1953, two days at the Institute of Nuclear Studies in Chicago in November, two days in January 1954 at the University of Minnesota, two days in February at the University of Indiana, two days in April at the University of Iowa, and two days in May at Purdue University. Because of teaching obligations, the meetings were always on Saturday and Sunday. These meetings were typically attended by approximately 20 people, usually including Cole, Robert Haxby (Purdue), Jones, Johnston, Laslett, Morton Hamermesh (Argonne), John Livingood (Argonne), Powell, Rohrlich, Symon, Terwilliger, Wright, Zaffarano, Kerst, and James N. Snyder (Illinois). Occasional visitors to the meetings included Abraham Taub, Enrico Fermi, Herbert Anderson, Keith Brueckner, Ken Watson, Josef M. Jauch, Rollefson, Crane, Alan Mitchell, and Kruger. The work from mid-1953 to mid-1954 was aided by a National Science Foundation grant of US$21,800, and by generous leaves of absence, travel, and facility support from the universities.

The biggest effort in studies over this period was in theoretical work (both analytical and computational) on orbits, although other topics studied included power supply ripple (at the BNL Cosmotron), linear accelerator injector design, frequency modulation of rf cavities, and magnetic remnant field effects. Jones and Terwilliger constructed an electromechanical device for demonstrating strong focusing and observing nonlinear effects; it was basically a galvanometer structure, with the coil driven by a current proportional to its angular displacement from the center and its field driven by a square-wave ac current. Crane had earlier noted that nonlinearities could lead to a kind of stability around relatively large amplitude oscillations, and much study went into understanding this "Craniac" motion, including its demonstration on the Michigan electromechanical model. Digital computers were just becoming available, with the ILLIAC at the University of Illinois in the forefront, and members of the MAC group were early and effective users of this facility.

A second summer session was held in Madison from mid-June to August 14, 1954. The effect of space charge, the coupling between the rf and orbital motion, and the development of the "smooth approximation" (by Symon) were some of the topics studied. One interesting study was the concept of thin-lip magnetic pole designs, where a thin-iron/low-field/large-aperture magnet pole would accommodate the beam at injection; the beam would then move into a narrower-gap, high-field region as it was accelerated and its betatron oscillation amplitudes adiabatically damped. Brueckner, Robert Sachs, and Watson kept the group informed about high-energy particle physics, and Courant and Snyder from Brookhaven again joined the group. Laslett and Haxby spent the summer at Brookhaven working with the accelerator development group there and learning about their magnetic measurement devices. Wright went to Berkeley to work with the Bevatron group.

Francis T. Cole (1925–1994)

Francis T. (Frank) Cole was perhaps the most active member of the MURA accelerator physics group until he left to go to Berkeley after the MURA proposals were turned down. Without doubt he — more than anyone else — created the spirit of MURA. He was a theorist and an accelerator designer who was involved with every one of MURA's machines.

Frank was born in New York City, earned his bachelor's degree at Oberlin, and his Ph.D. at Cornell University in 1952. He joined the University of Iowa's Physics Department faculty following his Ph.D. in 1952, a position he held until he moved to the MURA group in 1955. As one of the MAC physicists who were members of the 1953 Summer Study group at Brookhaven and later at Madison, he was a regular participant in the MAC and MURA weekend meetings during 1953–1955.

In January 1955, he moved to Urbana, Illinois where he was a member of the first full-time MURA working group. One of his first jobs there was to compute and fix the parameters for the Radial Sector Model, to be built at Michigan, with its magnets constructed at Purdue. Following the summer workshop at Michigan, he remained at Illinois through the following academic year (to which he referred in his work "Oh Camelot!" as the "glorious year"), then moving to Madison with Kerst in June 1956.

Frank was a mainstay of the MURA theory group throughout the life of the MURA organization, until the end of its accelerator program in 1964. More than anyone, he was the person who developed and wrote most of the many "proposals" from MURA.

In 1964, he moved to Lawrence Berkeley Laboratory, where he was involved with the "200 BeV Study," headed by Edward Lofgren. In 1967, when Illinois was selected as the site for the next step in accelerators, he moved there as one of the founders of the National Accelerator Laboratory (now known as Fermilab). He served as Associate Director during construction. Then, together with Don Young, he turned to operation of the facility, in which role he trained all the early operators of the Fermilab Accelerator, many of whom are now amongst its senior technical people. He remained at Fermilab until his retirement in 1992. Frank was involved in the details of the designs of the many MURA models and large (multi-GeV) accelerator proposals. Besides his many works and publications of technical details of accelerator design, one of his most memorable works is his unpublished autobiographical history of MURA, "Oh Camelot!", which was widely read by accelerator scientists. This work was a stimulus and provided the very basis for this book.

Frank married Rosemary Bolton in 1956. He adopted two children from Rosemary's first marriage. Together they had two more children. Frank had many outside interests, particularly in education. He was a member, then chairperson for many years, of the Wheaton, Illinois School Board, and the College of Dupage.

Lawrence Jackson Laslett (1913–1993)

Lawrence Jackson Laslett was the outstanding "house theorist" at MURA. He could calculate anything: often both analytically and numerically. No problem was beyond him. Besides being a competent workman in theoretical physics, he was also imaginative. As a participant in MURA he was a leader in the development of alternating gradient focusing theory, the concept of colliding beams, and nonlinear orbit theory. Laslett was also a pioneer in the use of digital computations in orbit and field calculation. With James Snyder and Richard Christian, he did the first digital computation of magnetic fields. By studying the fine details of mappings, he was the first to show by digital computation that there is a real phenomenon of chaos.

Laslett was born in Boston, Massachusetts and raised in Pasadena, California. He completed his undergraduate education at the California Institute of Technology and did his graduate work at Berkeley under Ernest Lawrence. He was a key figure in the development of the first cyclotrons, building, in Copenhagen, with Sten von Friesen, the first European cyclotron.

As a professor at Iowa State University, Laslett produced significant, copious, and meticulous contributions to weak focusing synchrotron design. Soon, however, MURA became his major activity. He served the government in many ways. During World War II, he worked at the MIT Radiation Laboratory on the development of airborne radar. He was a member of the Office of Naval Research in Washington (1952–1953) and later in London (1960–1961), and he was the first head of the high-energy physics branch of AEC (1961–1963).

In 1963, Laslett went to Lawrence Berkeley Laboratory, where he worked for the rest of his life. Here he continued to be a leading figure in accelerator design and theory, making major contributions to the design of the Electron Ring Accelerator and to the Heavy Ion Fusion Accelerator program and, most importantly, to the subject of collective instabilities.

Laslett's contributions and influence were immense. He was exceedingly quiet, but quietly shared his insights, work, and perspective, so that a host of people in the field revere him as a mentor. He was a joyous person, always eager to work and to discuss physics.

He was exceedingly modest and just one example, in the words of Dieter Mohl, shows that side of Jackson: "I remember what happened, when there was an effort to persuade him to go to the 1971 Accelerator Conference in Geneva. He lifted up his feet and showed the holes in the bottom of his shoes (stockings included), saying he could not go to Geneva like that. Time to buy new shoes, he suggested, but that would be time lost for physics. This argument convinced the whole group."

As Mohl remembered, with thoughts that all of us who knew Jackson share: "He is no doubt one of the most competent, gentle and heart-warming accelerator physicists I met. If there is a good God and a heaven, Jackson will surely have a small computer terminal room where he will sit from 11 p.m. to 1 p.m. next day solving problems."

Robert O. Haxby (1912–1972)

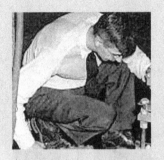

Robert O. Haxby was the MURA Laboratory's lead experimenter, and made great contributions to its programs. He built the magnets for the three electron models at MURA, and took part in the design of other components.

Bob Haxby got his Ph.D. in Nuclear Physics under John Williams at the University of Minnesota in 1939. His thesis was entitled "The Angular Distributions of the Disintegration Products of Light Elements Bombarded by Protons and Deuterons." Before World War II, he worked at Sperry Rand, where he was one of the inventors of the "Radar Speed Detector," still used by police to detect speeders. In 1945, he went to Purdue University as an associate professor, and there he built a 300 MeV electron synchrotron for studying meson production. He was promoted to full professor at Purdue in 1951. He took part in the early deliberations at MAC and MURA, and then, with the help of Ed Rowe, built the magnets for the Radial Sector (Michigan) Model. In 1956, he moved to the MURA Laboratory, where he supervised the completion, installation, and measurement of the magnets for the Spiral Sector (Illinois) Model. He then undertook the design, construction, measurement, and correction of the 50 MeV Model magnets. After the initial testing of this model in the winter of 1959–1960, he returned to Purdue, where he pursued experiments in particle physics with his fellow professors. In the mid-1960s, he moved to Iowa State University. He died of a heart attack while attending a meeting at the ZGS at Argonne in January 1972.

Bob and his wife, Mona, were warm and congenial people, and contributed greatly to the social life at MURA. While MURA was growing in size, from 1956 to 1963, Mona periodically hosted teas for the wives of MURA physicists to help them accommodate to their new environment.

3.2. THE INVENTION OF FFAG

Our story begins with the invention of alternating gradient focusing by Courant and Snyder at Brookhaven National Laboratory (BNL) [Courant, 1952], and independently by Christofilos [Christofilos, 1950] in Greece. A natural extension leads to what are now called fixed field alternating gradient (FFAG) accelerators, invented by K. R. Symon at MURA [Symon, 1954; Kerst, 1956] in the summer of 1954, and independently by T. Ohkawa [Ohkawa, 1954] in Japan, who was invited to join the MURA group in 1955, and also by Lee Haworth and H. S. Snyder [Livingstone, 1962] at Brookhaven and A. A. Kolomenskij [Kolomenskij, 1958] in the Soviet Union. In an FFAG accelerator the magnetic guide field is fixed, i.e., constant in time. The field is weak at the inner radius and strong at the outer radius,

Kent M. Terwilliger (1924–1989)

Kent M. Terwilliger was one of the small group of physicists who became involved with MAC and MURA early on, and became known as one of its most accomplished experimentalists. He and Lawrence Jones built the very first FFAG accelerator: the Radial Sector Model. He went on to be involved with many MURA projects during the next ten years.

Kent was born in San Jose, California and grew up in the Los Angeles area. During World War II, he served as a second lieutenant in the US army in Europe. He completed his undergraduate degree at Caltech in 1949, and his Ph.D. at the University of California, Berkeley in 1952, where he collaborated with Jones in studying photoneutron production and photon absorption by nuclei between 20 and 300 MeV in experiments at the UC Radiation Laboratory's then-new 320 MeV electron synchrotron and the 70 MeV electron synchrotron at the University of California, San Francisco medical campus. With Jones, he joined the University of Michigan faculty following receipt of his Ph.D. in 1952, and there they worked on H. R. Crane's new electron synchrotron and on nuclear physics research facilities at the 1930s cyclotron.

Kent and Jones joined the MAC group in 1953. Kent participated in the first summer workshops at Brookhaven and Madison, and continued his intensive involvement with MURA through the 1950s and early 1960s, including a year in residence at Madison (1956–1957), most summers spent at Madison, and frequent subsequent visits. With Jones, he built the first FFAG accelerator, the Radial Sector (Michigan) Model, and with it carried out pioneering studies of betatron oscillation stability and the dynamics of rf acceleration and phase displacement. He invented (at the Michigan synchrotron) the "rf knockout" means of measuring betatron oscillation frequencies and also the concept of orbit dynamics in which particles of different energies would have the same radius at specific azimuths in FFAG accelerators and/or in AG storage rings, in order to improve the colliding beam luminosity. He was a very sensible and wise physicist; it is recalled that sometimes when Kerst had a novel concept, he would say, "Let's see if I can get this past Terwilliger."

Following his MURA involvement, Kent worked with Donald Meyer at Michigan in the design of spark chambers and their subsequent use in experiments at the Argonne ZGS. At Argonne, he played a significant role in the establishment of the first "users organization," and was also an early member of the AEC (later DOE) High Energy Physics Advisory Panel. Later he also became involved with Alan Krisch's Michigan group in developing spin-polarized beams at the Argonne ZGS, the Brookhaven AGS, and the Indiana University Cyclotron Laboratory Cooler Ring, as well as in related experiments. He remained an active and productive member of the Michigan physics faculty until his untimely death from cancer.

Kent and his wife, Doris, had four sons; they enjoyed camping, picnics, hikes, fishing, playing bridge, and other social activities. Jones recalls him as a wonderful colleague and a delightful companion, and he and his family were great neighbors.

Tihiro Ohkawa (1928–)

Tihiro Ohkawa was one of the most imaginative and creative persons at MURA. Not only did he invent the concept of FFAG independently of the MURA physicists, but he also created the concept of the "two-way" FFAG accelerator for colliding beams (discussed in Section 3.9).

Ohkawa's first exposure to physics was at the early age of 16, when he was recruited, during World War II, to help in Prof. Yoshio Nishina's cosmic ray laboratory. Subsequently he received his Bachelor's degree and his Doctor of Science degree from the University of Tokyo.

Because of this independent invention of the FFAG concept, he was recruited by the MURA group, first as a visitor and then as a long-term staff member. During his MURA years, he worked on many things, but most importantly, he developed the concept of the two-way accelerator.

Subsequent to that period, Ohkawa spent some time at CERN in Europe, but then most of his years were spent doing plasma physics at General Atomics. During that time he was also a professor at the University of California, San Diego.

He has worked, most productively, in the various fields of nuclear physics, fusion energy and plasma physics, plasma processing and biotechnology. Besides his major contribution to accelerators he has, in fusion and plasma physics, contributed experimentally to the important problem of plasma transport and on the noninductive methods of generating current in toroidal plasma.

Ohkawa's work has led to over a hundred patents in plasma devices and biotechnology. He initiated the US–Japan fusion cooperation program and cofounded Archimedes Technology for nuclear waste disposal. He is a recipient of the Maxwell Prize from the American Physical Society. He was known as being a "most imaginative and creative person," and that characteristic extended far beyond his MURA days.

At the present time Ohkawa is at Archimedes Technology. He is interested in all sorts of inventions, and has even been characterized as an "invention junkie." His hobbies include skiing, tennis and golf. He is known for his debonair ways.

thus accommodating all orbits from injection to final energy. Focusing is provided by an alternating gradient. Section 3.3 contains an extensive discussion on the various FFAG configurations, including derivations of the formulas relating the various accelerator and orbit parameters.

The chief advantage of an FFAG accelerator is that the transverse motion in the magnetic guide field and the longitudinal motion under the accelerating fields are very nearly completely separated. As a result many accelerating methods may be

Fig. 3.2. A close-up photograph of the Radial Sector Model with one magnet removed, showing the vacuum tank, the support table, and details of the magnets. The insulated junction of the two halves of the vacuum tank, across which the rf accelerating voltage would be applied, is clearly shown here. Note that the supporting table is also joined with an insulated structure, necessary for betatron acceleration; the betatron core is also clearly apparent.

used, including betatron acceleration and a wide variety of radio-frequency accelerating schemes. Beam stacking schemes to be discussed later provide the possibility of producing intense beams. Accelerator experiments are also greatly simplified when the guide field is decoupled from the accelerating process.

The configuration initially proposed is called a radial sector FFAG accelerator [Symon, 1954]. Figures 1.1 and 3.2 show the electron model built by Jones and Terwilliger at the University of Michigan and moved to the MURA laboratory in Madison, Wisconsin in 1956. Figure 3.2 shows the finished machine with one magnet removed; Fig. 3.3 shows one of the magnets. Donald Kerst invented the spiral sector configuration [Kerst, 1956]. An electron model completed in 1957 is shown in Fig. 1.2.

Relevant Dynamical Parameters and Relationships

It may be useful to spell out some of the relevant dynamical parameters and relationships. In an FFAG machine of the radial sector variety, there would be N identical sectors, with each sector containing a radially focusing (positive curvature) and a radially defocusing (negative curvature) magnet. For each magnet, we

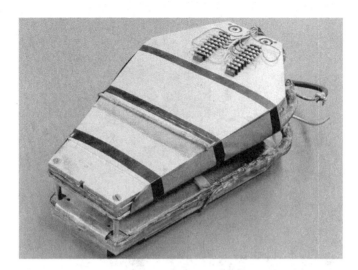

Fig. 3.3. A radial-focusing magnet from the Radial Sector Model; the pole-face windings are visible on the lower pole.

may define $\psi = n^{1/2}\,\theta = (B'/B\rho)^{1/2}\,s$. Here $n = |(\rho/B)\,(dB/dr)|$, $B' = dB/dr$, and $s = \rho\theta$ — the arc length of the particle within each magnet, with ρ the radius of curvature and θ the angle through which the particle is bent. [Note that $k = (r/B)\,(dB/dr)$; hence the gradient, dB/dr increases with r for constant k.] With ψ_1 for the positive curvature magnet and ψ_2 for the negative curvature magnet, the betatron oscillation phase advance, σ, per sector for the horizontal (r) and vertical (z) oscillations can be written as: $\cos\sigma_r = \cos\psi_1\cosh\psi_2$ and $\cos\sigma_z = \cos\psi_2\cosh\psi_1$. Of course, this is a simplified approximation, as the magnet edge focusing and the effects of straight sections between magnets are not included. A relevant representation of the design parameter space is the so-called "necktie" diagram of ψ_1 versus ψ_2, where the necktie-shaped region, bounded by $-1 < \cos\sigma_r < +1$ and $-1 < \cos\sigma_z < +1$, sets the boundary of the parameter space where the horizontal and vertical oscillations are stable. This is illustrated in Fig. 3.4.

A useful simplifying assumption about a radial sector FFAG magnetic field is that it is *scaling*, in the following sense. In terms of cylindrical coordinates r, θ, z, where z is the vertical coordinate, a scaling magnetic field has the form

$$\mathbf{B}(r,\theta,z) = \left(\frac{r}{r_0}\right)^k \mathbf{B}_0\left(\theta,\frac{z}{r}\right), \qquad (3.1)$$

where $\mathbf{B}_0$ is the field at a reference radius r_0, and the constant k is the mean field index. An equivalent definition is that at corresponding points that lie along any

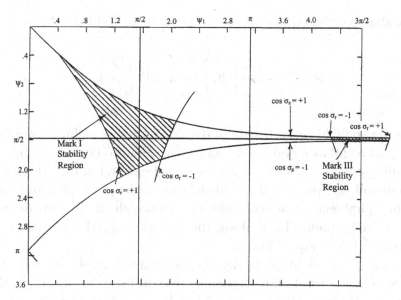

Fig. 3.4. The "necktie" stability diagram for a radial sector FFAG accelerator. Here ψ_1 and ψ_2 are the phase advance in radians of the radial and vertical oscillations within the radially focusing and the vertically focusing magnets of a sector (as defined in the text). The Mark I stability region, shaded, is bounded by $\cos \sigma_r = \pm 1$ and $\cos \sigma_z = \pm 1$; as is the "patch" (at the far right) stability region for Mark III. Here, $r = x + r_0$.

straight line through the origin, the magnetic field has the same direction and a magnitude which scales as r^k. If the magnetic field scales, then the particle orbits — both the equilibrium orbits and the betatron oscillations about those orbits — also scale in the sense that if any orbit at energy E_0 passes through a point at radius r_0, then there is a corresponding orbit at energy E that passes in the same direction through the corresponding point (on a straight line through the origin) at radius r, where

$$\frac{p}{p_0} = \left(\frac{r}{r_0}\right)^{k+1}, \tag{3.2}$$

p being the momentum corresponding to energy E. For a scaling FFAG field, if an orbit is stable at some energy E, then the corresponding orbits at all energies are stable. In particular, the linear horizontal and vertical betatron tunes ν_x, ν_z (betatron oscillations per revolution) are the same at all energies; the tunes determine the linear stability of the betatron oscillations. For nonscaling fields, it is much harder to guarantee the stability of the orbits at all energies. Discussions on scaling can be found in many MURA reports and papers [Cole, 1959; Laslett, 1956; Symon, 1955B].

For a scaling spiral sector accelerator, the magnetic field has the form

$$\mathbf{B}(r,\theta,z) = \left(\frac{r}{r_0}\right)^k \mathbf{B}_0\left(\theta - \tan\varsigma \ln\frac{r}{r_0}, \frac{z}{r}\right), \qquad (3.3)$$

where ς is the angle between the spiral direction and the radius. Equivalently, the magnetic field scales as r^k and has the same direction everywhere along a spiral of corresponding points which lies on a cone with the center at the origin, and where the projection of the spiral on the median plane makes an angle ς with the radius. For a scaling spiral sector magnetic field, the orbits scale in the sense defined above for corresponding points lying along the specified spiral, with the radius and momentum related by Eq. (3.2).

An FFAG magnetic field can be designed so that the radial variation of the field is such as to keep the revolution frequency constant [Symon, 1956A, p. 1856; Kerst, 1955]. We then have a nonscaling FFAG cyclotron, which can be designed with either radial or spiral sectors. In a cyclotron, the number of horizontal and vertical betatron oscillations per revolution varies with energy, but can be kept below unity; so beam-destroying resonances can be avoided. A cyclotron similar to the radial sector FFAG cyclotron was invented previously by L. H. Thomas [Thomas, 1938]. In scaling FFAGs, the revolution frequency, equal to the ratio of the speed to the circumference, is not constant, but depends on energy. Typically, the circumference increases with energy, but as the particle becomes relativistic, the speed changes very little. At some energy, the *transition energy*, the change in speed is just offset by the change in energy, so that the revolution frequency remains constant as the energy is changed. The result is no phase focusing or stability at that energy. Further, the phase of the rf at which the beam can be accelerated changes from where the accelerating frequency is increasing with time (below transition) to where it is decreasing with time (above transition).

3.3. MURA STUDIES

The Michigan Working Group, Autumn 1954

After the Madison summer session of the MAC group and Symon's invention of FFAG, Kerst decided that a more intense, ongoing program was desirable, although the periodic, larger weekend meetings would still be held. As Jones, Terwilliger, and Crane were at the University of Michigan in Ann Arbor, and Symon was at Wayne University in Detroit (and lived in the northwest Detroit suburb of Walnut Lake, within commuting distance of Ann Arbor), Kerst suggested that weekly meetings be

Fig. 3.5. The Michigan Working Group, autumn 1954. Seated, R-L: Lawrence Jones, Keith Symon, Kent Terwilliger, Donald Kerst, H. Richard Crane; at the blackboard: L. Jackson Laslett.

held at the University of Michigan. (Interestingly, Wayne was then a city university under the Detroit Board of Education, a unique situation; it is now Wayne State University.) Each week, Kerst and Jackson Laslett came to Ann Arbor by overnight train for two or three days (Wednesday–Friday), and the Michigan Physics Department provided an office for them, and they were joined by Symon, who drove to Ann Arbor from his home. Crane also frequently joined in these discussions, and Bob Haxby, Mel Ferentz (Argonne), Frank Cole, and Jim Snyder came to Ann Arbor occasionally to work with the group. Symon remembers an early meeting where Crane led a discussion on what to call the new magnetic configuration. It was at that time that the term "fixed field alternating gradient" (FFAG) was coined.

During this period, funds from the Atomic Energy Commission (AEC) were not forthcoming, and the group turned to the Office of Naval Research and the National Science Foundation. Most of the travel, experimental, and computational costs were borne by the universities. At the end of September 1954, the Organizing Committee brought about the incorporation of the group into MURA, an Illinois company.

The original plan for this Michigan study group, as described to the MURA Board of Directors, was to commence engineering studies, and specifically to examine the next higher energy for accelerators beyond the Brookhaven AGS and CERN PS 30 GeV range. Some studies were to be undertaken on means of

James N. Snyder (1923–1985)

James N. Snyder made important contributions to the MURA computing effort. In fact, he was the person who led in developing digital computation programs which allowed MURA, for the very first time, to move into numerical accelerator study and design.

Snyder received his Ph.D. in Physics from Harvard University in 1949. He joined the faculty of physics at the University of Illinois at Urbana-Champaign in 1949, reaching the rank of professor in 1958. He worked in the Digital Computer Laboratory, where he spent some time working with MURA people under the leadership of Don Kerst, then a professor at Illinois. The Digital Computer Laboratory later became the Department of Computer Science, and in 1964, Snyder was appointed Associate Head of Computer Science, and in 1970, he became Head of the Department. In the early 1960s, he pioneered the use of the ILLIAC computer for the processing of high-energy data. He developed operating systems for the IBM 650 and 7090/7094 computers.

With the invention of FFAG in 1954, and the highly nonlinear forces which act on circulating particles in that configuration, numerical computation of particle orbits became important. Snyder took the lead in developing programs for such computations. Many of the group which would later become MURA joined Kerst and Snyder at Illinois for various periods of time to help with computations on ILLIAC. Computing on ILLIAC was not easy. Memory was only about 100 words; computations had to be performed in fixed point (therefore at every stage in the computation the number could not be too large or too small); and the input/output was slow (paper tapes, which were then fed into a teletype machine, or the reverse process). In short, a real master was required, and that is exactly what Snyder was. These were among the first numerical computations of particle orbits in accelerators. Richard Christian from Los Alamos joined the group and taught them, particularly Snyder, how to compute magnetic fields in complicated geometries.

After the MURA group assembled at Madison and rented an IBM 704 computer, Snyder came up from Illinois to head the computer work. He ran a tight ship, with his thumb on all computing and all programming. IBM sent two programmers to Madison to learn scientific programming from Snyder and Christian. They developed a great variety of programs for calculating magnetic fields and particle orbits.

Snyder married Betty Jane Cooper in 1944. They had a son, James Newton Snyder. Throughout his career, Snyder spent summers on particle physics research. Fred Mills recalls meeting with him at Brookhaven one summer in the 1970s. He asked Jim, "Why do you concentrate on computers instead of physics research?" Jim answered, "It's better to be a big fish in a small pond than to be a small fish in a big pond." It is interesting to see how times change.

automatic or servocontrol of magnet misalignment, determined by beam sensors. Because of the very large circumference of higher-energy accelerators, very careful alignment would be required in order to have the desired small aperture, and these servo studies were the first topic when the study group met. However, the interests of this group were so fundamental, and general, that engineering occupied very little time.

Most of the discussions centered on Symon's recent invention of FFAG. The relevant orbit calculations, nonlinearity considerations, etc. occupied much of the time. There were also efforts to think of ways of reducing the "circumference factor" — defined as the circumference divided by 2π times the minimum radius of curvature. Different concepts were given different designations. For example, Mark I was the original, "standard" FFAG. Mark Ia had magnets with alternating field directions, but of equal size, however with the magnetic field strength in the outward-bending ("negative," vertically focusing and radially defocusing) magnets weaker than in the central-bending ("positive," radially focusing and vertically defocusing) magnets. Mark Ib had magnetic fields of equal strength, but with the positive magnets longer (in azimuth) than the negative magnets. The Mark I designs were all "scaling," meaning that the equilibrium orbits of different energies were identical except for a radial scale factor. The fields increased with the radius as a power law: $B = B_0(r/r_0)^k$. Hence the numbers of betatron oscillations per revolution, v_r and v_z were constant. However, a constant k corresponded to a radially increasing gradient, dB/dr, thus leading to essential nonlinear terms in the betatron oscillations.

Mark II was a nonscaling concept, where the negative magnets were wider and the positive magnets narrower at injection (the inner radius) and tapered so that the negative magnets had a shorter azimuthal length, with a higher k value at full energy, resulting in a smaller circumference factor than Mark I. This nonscaling geometry enabled a larger aperture and easier stability for the injection orbits than the Mark I designs, but had many other problems. For example, the edge focusing of the magnets, tapered with the radius as noted above, was vertically defocusing, and resulted in a circumference factor hardly better than for Mark I. Another major challenge was to keep the v values independent of energy in such a nonscaling structure. An advantage of scaling is that keeping v constant makes it easier to avoid resonances. The group believed in those years that the betatron oscillation frequencies, v_x and v_y must be kept constant throughout the acceleration cycle to avoid crossing a resonance and thus losing the beam.

In Mark III, the positive magnets were much longer than the negative magnets, so that the radial betatron oscillations experienced a phase advance of over a half betatron wavelength in one sector (a positive plus a negative magnet). The vertical focusing, though still in the stability region, was quite weak. In a graph of the focusing strength of the positive magnets versus the focusing strength of the negative magnets, the region of stability is a diagonal strip, which was called the "necktie" (Fig. 3.4). This central region of stability includes orbits with less than half a betatron oscillation per sector, both radially and vertically. There are also regions off the necktie with more than half an oscillation (phase shift of over 180°) of horizontal and/or vertical oscillations, which were called "buttons" or "patches." Mark III was a design on one of these patches, where the much longer positive, radially focusing magnets had the same field and k as the shorter negative magnets. It was a scaling geometry (as was Mark I). Its advantage was that it had a much better circumference factor than Mark I or Mark II; its disadvantage was that the vertical oscillation frequency was very sensitive to the field gradients (or focusing strengths).

Mark IV was also a nonscaling idea, where, in order to achieve a smaller circumference factor, all magnets had the same polarity but with opposite gradients. Hence the low-energy orbits crossed the high-energy orbits between magnets. Thus the maximum energy orbits were almost circular, and the injection orbits wildly oscillating. In all of these designs (especially Mark II and Mark IV), the orbit oscillation frequency calculations required consideration of the magnet edges. When a particle crosses a magnet edge (from high field to zero field, or from zero field to high field) at an angle other than 90°, the fringe field exerts a focusing (in one dimension) and defocusing (in the other dimension) force on the particle, with the focusing strength proportional to the tangent of the angle between the orbit and the normal to the magnet edge, i.e. the focal length of the magnet edge "lens" is given by $f = \rho \tan\theta$, where ρ is the radius of curvature (of the particle in the magnetic field) and θ is the angle between the normal to the magnet edge and the particle trajectory. One complication was the modification of the focusing effect of the edges in the realistic configuration with a finite magnet aperture and a corresponding slow change in the field, rather than a field which went from zero to maximum in a step function. This "soft edge" required a more sophisticated calculation for the effect on the oscillation frequencies than did the "hard edge."

Kerst, on one of his train trips to Michigan, had the brilliant idea of using this edge focusing to supply all of the vertical focusing by creating a spiral magnet design, Mark V. Here the magnets scale, but are all positive (radially focusing, positive k value), with either lower-field regions or zero-field spiraled straight sections between them. Of course, the edges are alternately focusing and defocusing, and with the gradient in the spiral magnet sectors, the net effect is stronger radial than vertical focusing (as in Mark I), but with vertical focusing still within the stability limits. This immediately led to designs with a circumference factor near 2, instead of about 5, as in Mark I.

Toward the end of the autumn, the idea of building an operating electron model FFAG accelerator evolved, and it was decided to have Jones and Terwilliger construct it at Michigan.

In those days, there was an annual meeting of the American Physical Society in Chicago (at the University of Chicago) every November. The MURA group arranged its general meeting to coincide with that meeting. I. I. Rabi, from Columbia, came to that meeting and spoke to a gathering of the MURA group and Argonne physicists, about the desire of AEC's General Advisory Committee (of which he was the chair) to establish Argonne Laboratory as an accelerator development lab, and he encouraged the MURA physicists to move to Argonne. None of the MURA group were interested. Argonne had a reputation for being totally involved in nuclear engineering; its activities were classified and entry to the laboratory required clearance and other red tape. And, of course, the AEC management contract for Argonne was held solely by the University of Chicago, with no other Midwestern universities involved. Rabi had been the organizing genius behind Brookhaven, and also played a major role in the creation of CERN, and this was the ideal of a regional laboratory, which the MURA group was seeking. But the AEC management was unenthusiastic about creating another national laboratory in the Midwest. It should be noted, however, that relationships with Argonne accelerator physicists had been, and continued to be, excellent. Morton Hammermesh, John Livingood, Edwin Crosbie, Melvin Ferentz, and — later — Lee Teng were often participants in and contributors to the MURA meetings and discussions. Courtenay Wright from the University of Chicago was also an early participant in the MURA activities.

The Illinois Working Group, Winter and Spring 1955

At the close of 1954, the Michigan weekly meetings ended, and the theoretical focus shifted to Urbana. During the winter and spring of 1955, Frank Cole moved to Urbana, Laslett was a regular visitor, and Jim Snyder (with the ILLIAC computer) became more involved. Nils Vogt Nilson from Norway also joined the group. It was there that the details of the parameters for the Radial Sector (Michigan) Model were worked out, and the magnets designed. There were frequent meetings at Urbana, which involved Jones, Terwilliger, Haxby, and others, including Edward Akeley, an older theoretical physicist from Purdue. At Michigan, the necessary hardware for the model was being built and assembled — the vacuum tank, power supplies, betatron acceleration core, injectors, etc. As they were completed later in the spring and during the summer, the magnets were brought (one at a time) to Michigan, the first ones by Kerst on the train on his Michigan visits. As earlier, the larger weekend general meetings continued almost every month — at Northwestern University in January, the University of Indiana in February, and the University of Minnesota in April.

A highlight of this period was the presentation of three 10-minute contributed papers on FFAG at an APS meeting (discussed further in Sec. 3.11).

The MURA Summer Study at Ann Arbor, 1955

With the Radial Sector (Michigan) Model under construction, it was decided to hold the 1955 Summer Study at the University of Michigan. A new nuclear engineering laboratory had just been completed on the North Campus of the university, which would later house a 2 MW reactor; however, at this time it was largely unoccupied, and so was available for both offices and meeting rooms for the group, and provided laboratory space in which the Michigan Model could be constructed. Besides the core group of Midwestern physicists (including Kerst, Cole, Norman Francis, Laslett, Jones, Lawrence Johnston, Haxby, Terwilliger), Andrew Sessler, Ernest Courant, David Judd, Felix Adler from Carnegie Tech, and foreign visitors Tihiro Ohkawa from Japan, Nils Vogt Nilsen from Norway, and Otto Frisch from England (famous as a pioneer in nuclear fission) joined the group (Figs. 3.6 and 3.7). Work on the construction of the Radial Sector (Michigan) Model continued.

Of course, the theorists continued working on various problems. The following list of topics is taken from Kerst's 1985 paper [Kerst, 1985].

- The study of tuning methods for, and observations of, orbital and betatron frequencies in the Michigan Model.
- Spiral sector (Mark V) magnet design problems and methods of introducing straight field-free sections into the spiral magnet.

Fig. 3.6. Members of the 1955 MURA Summer Study in Ann Arbor, L-R: Ernest Courant, Tihiro Ohkawa, Otto Frisch, and David Judd, by the unfinished Radial Sector Model.

Fig. 3.7. Members of the 1955 MURA Summer Study in Ann Arbor. L-R: Ernest Courant, Tihiro Ohkawa, David Judd, Nils Vogt-Nilsen, Kent Terwilliger, Felix Adler, and Otto Frisch.

- Radio frequency (rf) beam-handling issues; what happens to particles lost out of the phase-stable region?
- Could the rf phase of a cavity be automatically controlled by sensing the beam bunch as it rotates?
- What happens to betatron oscillations excited by the synchrotron (rf acceleration) gap?
- There were also studies of possible computer solutions to problems of nonlinear forces in two dimensions for all types of accelerators, and the introduction of magnet errors into the computer simulations, plus the problem of coding nonscaling machines, and beam extraction questions.

The Illinois Working Group, Academic Year 1955–1956

During the academic year 1955–1956, Kerst felt that he must continue the full-time resident group of physicists to work more intensively on the many concepts and problems that the MURA group had been studying. Primarily because of the ready availability of computer resources, this group was established at the University of Illinois. Included in the group were Laslett, Sessler, Vogt-Nilson, Snyder, Cole, Lloyd Fosdick, Symon, and Akeley (who commuted from Purdue). The problems of colliding beams (Sec. 3.9), and the topology of FFAG machines to achieve them, were studied in detail. Cole in "Oh Camelot!" [Cole, 1994] describes this as "the glorious year." The group commenced the design of an electron model (on the scale of the Michigan Model) that would employ spiral sectors (Mark V); this became known as the Spiral Sector (Illinois) Model, as the magnets and other hardware were made on the Illinois campus in the shops of Kerst's betatron laboratory (Sec. 3.5). Thomas Elfe and Frank Peterson did a large part of the technical work, and Peterson remained with the MURA group (later in Madison) for many years.

An important development of this period was the comprehensive Hamiltonian theory of longitudinal motion and rf acceleration by Symon and Sessler (Sec. 3.4), which was a critical component in the evolution of practical colliding beam concepts, beam stacking, phase displacement acceleration, etc., as well as a significant and practical contribution to the understanding of rf acceleration in synchrotrons in general.

Laslett collaborated with Snyder in the development of magnetic field relaxation methods; this work became even more important the following year, when there was a larger computer and a collaboration with Richard Christian, who joined the MURA group then, following his interaction with Kerst at Los Alamos.

The first general meeting in the fall of 1955 attracted approximately 70 high-energy and accelerator physicists from all parts of the country. The topics covered included problems in high-energy particle physics, the East Coast and West Coast accelerators, and the new FFAG accelerator possibilities, including the colliding beam concept. In 1956, a proposal was made to AEC for a pair of 25 GeV high-current FFAG spiral sector accelerators tangent to each other with a colliding beam capability incorporated.

During this period, funding was first obtained from NSF and the Office of Naval Research, and later (finally!) with help from AEC.

The above discussion was significantly aided by Kerst's 1985 Fermilab symposium paper [Kerst, 1986], by Cole's "Oh Camelot!" [Cole, 1994] and by minutes of the weekly meetings and general meetings during 1954.

3.4. THEORY OF RADIO FREQUENCY ACCELERATION

Keith Symon and Andrew Sessler worked out an analytic theory of the radio frequency acceleration process in FFAG accelerators [Symon, 1956B] and carried out numerical experiments to confirm it. The computer code to study rf acceleration was written by Jim Snyder for the ILLIAC.

Symon and Sessler were able to formulate the equations of motion during rf acceleration in Hamiltonian form. Jones and Terwilliger installed an rf system on the Radial Sector Model and did experiments confirming the theoretical predictions. One acceleration scheme they studied was based on a process they called

Richard Christian (1923–1988)

Richard Christian was the genius behind most of the field computation programs that marked the progress at MURA. Together with Snowdon and Young, he developed relaxation techniques to accelerate the convergence of numerical calculations of static magnetic and radio frequency electromagnetic fields defined on lattices or meshes. These techniques were widely copied at other laboratories.

Dick was born in Chicago and received his B.S. degree in 1947 from the Illinois Institute of Technology, where he studied Chemical Engineering and Physics, and his Ph.D. in Physics in 1951 from the University of California, Berkeley, under the direction of Robert Serber.

Before and after his studies at Berkeley, Dick was an employee at Los Alamos National Laboratory. He was particularly prized for his digital computing ability. While there, he married Marie Roybal of the San Ildefonso Pueblo, where they built a magnificent adobe house. Keith Symon remembers visiting the Christians at Los Alamos, where he first met Dick. He was particularly amazed at Dick's practice of carrying around a set of punched cards, which he would shuffle to get them in proper order for the particular problem he wanted to solve, then feed them into the computer, which always put out the required answer.

Impressed by Dick's computing skills, MURA invited him in 1955 to join the group at Madison. At MURA, he exploited relaxation methods to calculate complex magnetic fields in many accelerator geometries. FFAG fields are highly nonlinear, and Dick played a leading role in learning how to integrate orbital differential equations numerically. He showed how to use nonrectangular meshes, to do curved boundaries, to calculate the effects of the exciting currents (with collaboration from Jackson Laslett and Stanley C. Snowdon on the physics), and to calculate the effects of saturation in the magnet steel. Later, he extended relaxation methods to wave equations and founded the methods still in use for calculating the radio frequency fields in linear accelerators. At MURA, he designed the copper–iron magnet for the MURA 30-inch bubble chamber used at the ZGS and later at Fermilab. The 30-inch magnet provided a field strength of 30 kilogauss, with unprecedented uniformity.

Toward the end of his service with MURA, Dick was offered a position on the physics faculty at Purdue University. He accepted and in 1964 joined the Purdue faculty as Associate Professor of Physics. While at Purdue, he developed an interest in plasma physics. He devoted his efforts to a mathematical model of turbulence for plasmas and fluids. This led him into the field of chaos and catastrophe theory. For some two decades he trained a cadre of students, who went out into the world to work in these disciplines. Dick had a keen interest in the application of physics to many areas of knowledge. For example, a casual conversation with Arnold Tubis concerning the pitch clarity of the orchestral timpani led to the publication of a widely referenced paper about the effects of air loading on the complex modal frequencies of the timpani.

Unfortunately, Richard Christian's career was cut short by a stroke on March 1, 1988.

"beam stacking." Since the space charge limit is smaller at low energies, one can increase the accelerated beam intensity by injecting successive beams at the space charge limit at injection and accelerating them to an intermediate energy. Space charge is the mutual Coulomb repulsion of the beam particles. It leads to a change in the tune of the betatron oscillations and a consequent crossing of a destabilizing resonance above some particle density, called the "space charge limit." After an intense beam has been stacked, it can then be accelerated to a higher energy.

Figure 3.8 shows the results of a numerical simulation of an rf acceleration process in which the rf and voltage are fixed. Once per revolution one plots a point at the particle energy and the rf phase when the particle arrives at the accelerating gap. There is a fixed point at phase π, energy 500 MeV, where the rf is nine times the revolution frequency, and another at 814 MeV, where the rf is ten times the

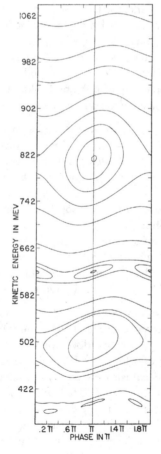

Fig. 3.8. A theoretical study of longitudinal phase space as generated by a single rf cavity. The ordinate represents kinetic energy and the abscissa, rf phase. The various stable (and unstable) harmonics are apparent.

revolution frequency. Both points are surrounded by trapping regions where the points lie on closed curves surrounding the fixed points.

An early acceleration scheme, considered by Symon, proposed modulating the rf so that the trapping regions (as in Fig. 3.8) move upward in energy. If the harmonic number, i.e., the ratio of the rf to the revolution frequency, is high, there will be a large number of closely spaced trapping regions. As each moves past the injection energy, a bunch of particles is injected. The trapped particles are accelerated until they reach the output energy. By analogy with the device used by farmers to lift grain into a storage bin, Symon proposed calling this scheme a "bucket lift." The trapping regions came to be called "buckets." Although no bucket lift accelerator was ever built, the name "buckets" came into common use and is Symon's contribution to accelerator terminology.

It was first observed numerically that if a bucket is moved upward in energy the surrounding phase space moves down. In fact this observation was made on the very first computer test run. Loading with only 11 particles, it was observed that 8 moved in energy the opposite way from what was expected! Theoretical understanding came quickly: if particles are near, but outside of, a bucket, as is always the case for particles previously accelerated and stacked at some energy, they will be phase-displaced to a lower energy by a bucket moving upward. This term became a second contribution to accelerator terminology.

With high rf voltages, stochastic phenomena were observed (now called "chaos") near the boundaries of a bucket, as shown in Fig. 3.9. On the hypothesis that stochastic phenomena occur when bucket boundaries overlap, they ran a case with two nearby rf frequencies with voltages such that the predicted buckets would overlap. The result shown in Fig. 3.10 gives totally chaotic orbits. The solid curves are the predicted bucket boundaries.

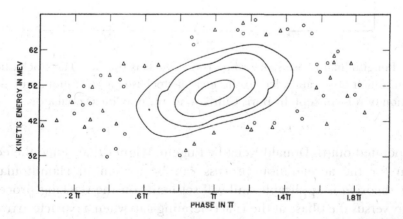

Fig. 3.9. Longitudinal phase space near a resonant harmonic of a high-voltage rf bucket (stable region). One can see the stable orbits and also the scattered points at larger amplitude.

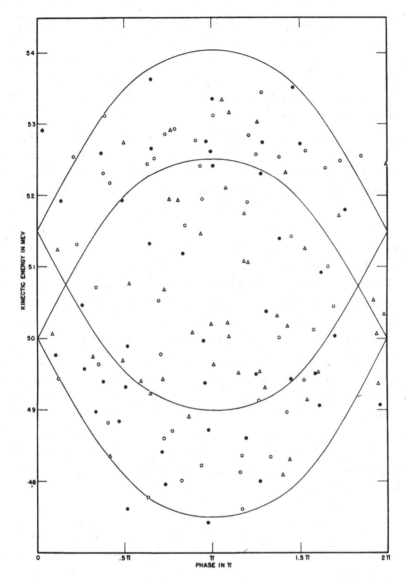

Fig. 3.10. Longitudinal phase space when two rf buckets overlap. The stochastic motion can be seen, which observation led to the idea of stochastic acceleration, and also to the generalization of this concept, by Boris Chirikov, to any non-linear oscillator.

It was pointed out to Donald Kerst by Eugene Wigner that, since the equations of motion for the acceleration process can be written in Hamiltonian form, Liouville's theorem is applicable and will set limits on the stacking process. If we plot energy versus the phase at the rf accelerating gap when a particle arrives there, the points in this phase space moving according to the equations of motion will move like an incompressible fluid (in two dimensions). This implies that the area

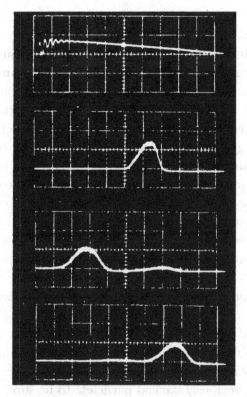

Fig. 3.11. Oscilloscope photo showing phase displacement using betatron detection. The energy is higher to the left, as the higher energy particles reach the target earlier. From top to bottom: (a) The betatron accelerating pulse. (b) The photomultiplier pulse of the beam striking the target with the rf starting frequency above the electron coasting frequency. (c) The rf starting frequency at the electron coasting frequency, so that most of the beam is rf accelerated and comes out earlier in the betatron acceleration pulse. (d) The rf starting frequency is below the coasting frequency; only a little of the beam is captured, and the rest is displaced to lower energy — thus it reaches the target later in the betatron pulse.

in phase space occupied by the stacked beam will be at least equal to the sum of the areas of the rf buckets that brought the particles to that energy. This sets a minimum on the energy spread of the stacked beam. Because the concept of Liouville's theorem is so physical, especially in a one-dimensional system, many members of the MURA group were soon thinking about rf acceleration.

Figure 3.11 shows the result of a phase displacement experiment on the Radial Sector Model [Terwilliger, 1957A]. In rf acceleration experiments like this, the beam is accelerated by a first betatron pulse to a starting energy (300 keV) for the experiment. An rf acceleration program is then executed. The resulting distribution in beam energy is then analyzed by a second betatron pulse, which accelerates the beam onto a target from which the beam intensity signal is plotted. The highest-energy parts of the beam are closest to the target and arrive first, followed by lower-energy beam. Hence, in the plots, the horizontal time scale is effectively an energy scale, with the energy increasing to the left. In the figure, the first trace is the second betatron pulse. The remaining traces show the beam arriving on target during the analyzing betatron pulse. In each case two complete rf modulating cycles were carried out before the analyzing betatron pulse. The second trace is the result when the rf is turned on at a value above the beam revolution frequency and modulated up to a higher final value. Thus the bucket starts at an energy above the beam energy and goes up to a higher energy; the result is the same as if the rf is not turned on at all. The time between the beginning of the second betatron pulse and the beam pulse represents the time required to accelerate the beam onto the target, and is therefore proportional to the energy difference between the initial coasting beam and the target. In the next trace, the turn-on frequency is synchronous with the beam, and most of the beam is captured and accelerated. In the last trace, the turn-on frequency is below the revolution frequency of the initial coasting beam. No beam is captured and instead the beam is phase-displaced downward in energy, hence arriving at the target about 24 µs later than in the second trace. Values of the phase displacement energy calculated from the experimental measurements agree roughly (within about 25%) with the theoretically predicted values.

Finally, it was shown that it would indeed be possible to achieve what Kerst desired: namely, to build up a stacked beam; i.e., to accelerate particles with rf while not having the rf destroy the previously stacked particles. To be sure, this is not 100% true, but it is sufficiently true for practical purposes. Symon's Hamiltonian formulation of the effect of rf on particles allowed study of particles either inside or outside the bucket. Sessler was able, using ILLIAC programs, to calculate orbits as well. As noted earlier, it was in Kerst's discussion of this observation with Wigner that the importance of Liouville's theorem had come up. With this concept and Symon's Hamiltonian, it did not take long for the MURA group to establish a complete understanding of stacking. The concept of stacking, of phase displacement, and other rf issues were studied experimentally with the Radial Sector (Michigan) Model during 1956, and later — in greater detail — with the Spiral Sector (Illinois) Model.

Figure 3.12 shows the results of an experiment in beam stacking carried out on the Radial Sector Model. The plot shows beam intensity versus energy with energy increasing to the left because of the way the energy was measured. The top plot is of the initial beam at an energy of 300 keV. The rf oscillator is then turned on at a frequency in resonance with the beam at 300 keV so as to trap it, the frequency is modulated upward to a value corresponding to an energy near 400 keV, and the rf voltage is turned off. The second plot shows the resulting beam distribution. Most of the original beam has been trapped and carried up in energy. Note also two features of the remaining beam. It is somewhat scattered in energy because this part of the beam was not in resonance with the rf. Its peak is also displaced slightly down in energy, illustrating phase displacement. The remaining plot shows the result

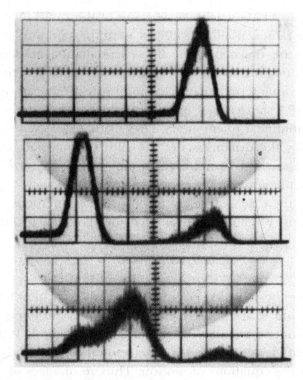

Fig. 3.12. A beam stacking experiment. Higher energy is toward the left. (a) Coasting beam (accelerated earlier, no rf). (b) One rf pulse, capturing the beam and accelerating it to a higher energy. (c) Four rf pulses (each the same as in (b)), illustrating the phase displacement of the higher-energy beam.

after four rf cycles. Most of the original beam has been picked up after four cycles. The phase displacement of the accelerated beam due to later rf cycles is clearly seen. If this were actual beam stacking, a new beam would be injected at the initial energy before each rf cycle. By studying these plots, one can determine the result of an actual beam stacking experiment.

The group studied, analytically and using numerical simulation, the capture of beam in an rf bucket, with particular attention to capture efficiency. They also studied acceleration through the transition energy, where the revolution frequency stops increasing as the energy increases and begins to decrease, and the bucket phases change abruptly.

3.5. NONLINEAR DYNAMICS

Prior to this time, beam physicists had only dealt with linear systems and short-time behavior, but the FFAG fields were quite nonlinear, and the interest was now in long-term stability. In FFAG machines, because of the rapid increase of the magnetic field with the radius, nonlinear effects are important and determine the stability limits, which are the maximum allowed oscillation amplitudes. Given the problem of maintaining beams for very long times, the MURA group started to ask questions about the long-term behavior of nonlinear dynamical systems.

As a result, MURA staff carried out extensive analytic, numerical, and experimental studies of nonlinear phenomena. They made extensive use of methods and results developed by the dynamicists of the 19th century and, also, of the more recent work of V. I. Arnold [Arnold, 1978] and Jurgen Moser [Moser, 1955] for studying nonlinear phenomena in celestial mechanics.

In particular, in the case of accelerator orbits, these methods predict that nonlinear resonances arise when the betatron oscillation frequencies satisfy the equation

$$n_x \upsilon_x \pm n_z \upsilon_z = m, \tag{3.4}$$

where v_x and v_z are the numbers of horizontal and vertical betatron oscillations per revolution, n_x and n_z are positive integers, and m is an integer. The plus sign corresponds to a sum resonance; the minus sign (with both v_x and v_z nonzero) corresponds to a difference resonance. The resonance (3.4) is driven by terms of order $n_x + n_z - 1$ in the equations of motion. Thus resonances of order 0 or 1 are linear resonances; higher-order resonances are driven by nonlinear terms. Sum resonances of order 3 and lower, and sometimes those of order 4, can produce instabilities even at infinitesimal amplitudes of oscillation. Figure 3.13 shows plots of the one-dimensional Hamiltonian corresponding to the third integral resonance $3v_x = m$. Here ε_x is the distance in tune space from the resonance, and β_x is the x amplitude function. On the resonance (case b, $\varepsilon_x = 0$) the motion is unstable at infinitesimal amplitudes. Below the resonance (case a) the resonance occurs at a finite amplitude. Above the resonance, if ε_x is below a threshold value depending on the amplitude of the term driving the resonance (case c), the resonance occurs at a finite amplitude. Above the resonance, if ε_x is above this threshold value, the motion is stable at all amplitudes. These results are due to the fact that the nonlinear terms generally drive the betatron frequency in a certain direction (assumed upward in Fig. 3.13), and the term that drives the resonance pulls the frequency toward the resonance. Higher-order resonances produce instabilities only at amplitudes above a finite amplitude threshold. Difference resonances do not cause instabilities, but they produce coupling between radial and vertical betatron oscillations.

Numerical orbit calculations confirmed these predictions. Experiments to check them were done on both models. Figure 3.14 is a contour plot showing beam intensity in the Radial Sector Model as a function of the number of radial oscillations per revolution plotted horizontally and the number of vertical oscillations per

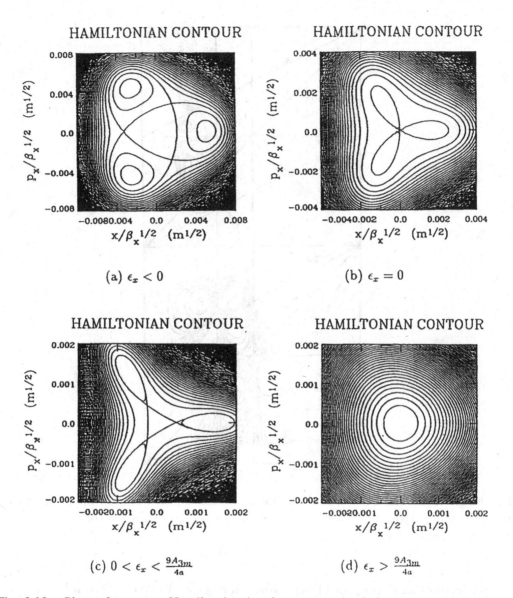

Fig. 3.13. Plots of constant Hamiltonian in phase space (scaled momentum vs. scaled position) for the resonance $3\nu_x = m$. Here A_{3m} is the amplitude of the term driving the resonance and $\varepsilon = \nu_x - m/3$ is the "distance" in tune from the resonance.

revolution plotted vertically. (See the later discussion in Sec. 3.6 of the way these experiments were done.) Resonances predicted by Eq. (3.4) lie along the straight lines shown. One can see the wide stop band along the linear resonance $\nu_x = 3$, as well as reductions in intensity along other linear and nonlinear resonances. Similar measurements made with the Spiral Sector Model (Fig. 3.15) also confirm the predictions of orbit theory.

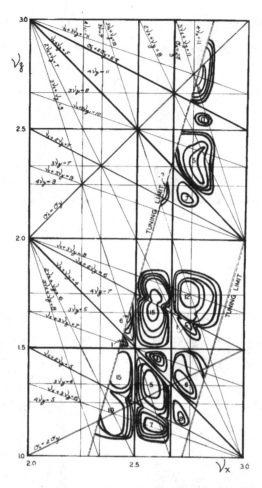

Fig. 3.14. Resonance survey for the Radial Sector Model. The contours represent (roughly) tunes of the same intensity, with the central area in each set of contours the highest intensity in that region of tunes.

Numerical calculations of FFAG orbits often showed apparently random behavior that was called "stochastic" behavior. Such behavior would now be called "chaotic." At first the MURA group was not sure whether these effects were real or artifacts of the numerical calculation. K. R. Symon and A. M. Sessler devised exactly canonical numerical algorithms to eliminate the possibility of nonphysical features of the algorithm. L. J. Laslett and Sessler made extensive checks to guard against round-off errors. They thus convinced themselves that these stochastic effects are real. The previous section gave some examples for longitudinal motion.

Colliding beams introduce the problem of maintaining beams for very long times, so the MURA group started to ask questions about the long-term behavior of nonlinear dynamical systems. Prior to this time, beam physicists had only dealt with

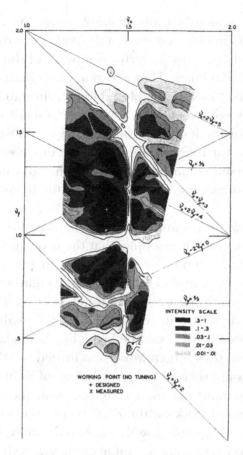

Fig. 3.15. Resonance survey for the Spiral Sector Model. The beam intensity is proportional to the darkness of the shading, as noted on the intensity scale.

linear systems and short-time behavior, but the FFAG fields were quite nonlinear, and the interest was now in long-term stability. The group did particle tracking for a few turns (since computers were not very powerful yet, in 1955), and then developed mapping techniques for longer runs (mapping techniques were much faster computationally). Maps relate the final values to former values by simple algebraic formulas. This was the first time that the concept of "maps" has been presented; now (in the 21st century) it is a tool that is used in essentially every laboratory studying beam dynamics. Quickly, the group saw that if the map was not exactly dynamical, i.e., preserving Poincare invariants (Liouville's theorem in one dimension), in just a few iterations one would obtain nonphysical results (such as damping in phase space). Thus Sessler and Laslett made what are now called symplectic maps. The maps that they employed were Taylor series developments, which were Hamiltonian and nonlinear, and corresponded to the actual accelerator through some order, but then had higher-order terms constructed to make them Hamiltonian. Thus some "real system," i.e., some physically realizable system, was described (although not necessarily the FFAG in question, except through some order).

With these maps and the most powerful computer of the time (ILLIAC), they could apply the map 50,000 times. They also ran it backward to be sure that it was free of truncation errors. Thus they explored long-term stability, and learned that one could design highly nonlinear fields (but linear at small amplitudes) that gave stable motion at least for the length of the run they could study. They never published any of this work, for they considered the results uninteresting — no new phenomena were observed — and, furthermore, they did not consider the runs long enough to make interesting statements about the long-term stability needed for colliders.

The MURA group was well aware of the deep nature of the questions that were being explored. They noted, for example, that the observed stability of planetary orbits in the solar system (with nonlinear effects due to gravitational perturbations caused by other planets, e.g., Jupiter) provided little comfort, since the operation of colliders required storing particles for a number of oscillations much greater than the age of the solar system in years (e.g., 4×10^9 oscillation cycles). Ernest Courant, who often visited MURA, said he had a brother-in-law who might help in these considerations, and thus Jurgen Moser was invited to MURA. He contributed much to the MURA group, such as the speculations of Kolmogorov, which boded well for these problems (and was even consistent with the numerical work cited above). Moser's subsequent work on the KAM theorem is widely known [Arnold, 1978; Michelotti, 1989; Guignard, 1989]. Many years later, Chirikov was able to develop a quantitative criterion that was quite consistent with the early observation at MURA (especially with rf) [Chirikov, 1979].

To summarize the MURA work on long-term stability: although long-term beam stability was not absolutely proven, the absence of long-term instability (at least for the length of time observed) motivated designing systems which would store beams for very long times.

3.6. THE RADIAL SECTOR MODEL

Michigan Working Group Discussions

During the autumn of 1954, when Kerst, Laslett, and Symon convened at Ann Arbor every week, it was decided that an electron model should be built to prove the FFAG principle and to test the various calculations of stability, resonances, etc. Jones and Terwilliger agreed to build the model at Ann Arbor; Frank Cole did most of the design work, and the magnets were built by Bob Haxby, with help from Ednor Rowe, a young Purdue physicist who subsequently spent his entire career at MURA and the

Physical Sciences Laboratory (PSL) at Wisconsin. It was decided that this would be the simplest and best-understood of the FFAG types — the "Mark 1b" with eight sectors of alternating radial positive and negative magnets (eight of each), each with the same field strength, but with the positive-bending magnets wider, and hence with a circumference factor of a bit over 5. It would utilize betatron acceleration to accelerate electrons to about 400–500 keV, from an injection energy of about 25–30 keV. The size was "tabletop" size, with a vacuum tank inner radius of 32 cm and an outer radius of 54 cm, and an overall diameter of about 1.55 m. This was known formally as the Radial Sector Model, and colloquially within the MURA group as the Michigan Model. Jones and Terwilliger had become educated in the hardware and technology of accelerators as graduate students at Berkeley, where — as part of their graduate research assistant responsibilities at the Radiation Laboratory — they became part of the operating crew of the 320 MeV electron synchrotron (on which they also did their thesis research).

Theoretical Design

The general idea for the machine was evolved in the late fall of 1954, but Cole worked out the detailed design after his arrival at Illinois in early 1955. The basic problem was that a significant part of the focusing came from the magnet edges and this was not yet understood in a realistic magnet with a finite aperture (called a "soft edge"). The group struggled several months with numerical hand calculations, using the Frieden and Marchand calculators; programs and the needed memory capacity did not yet exist for computer calculations (which a year later would have been much easier).

Work at Michigan: Table, Vacuum Tank, Injector, Betatron Core, etc.

The aluminum vacuum tank and table were constructed in halves, cemented onto plastic spacers (table), and bolted with insulating gaskets (vacuum tank) so that the induced acceleration voltage supplied by the betatron core would not be shorted out. The vacuum tank was machined by a Detroit tool-and-die firm that made the dies from which automobile body parts (fenders, etc.) were stamped. After the top had been heliarc-welded to the main tank, access ports were machined in the inner and outer radii of the tank for the injector, measurement probes, and other devices. Unfortunately, these resulted in vacuum problems (since the seatings for the ports required machining across the welds), with the consequence that the vacuum was no better than about 10^{-6} torr, so that the lifetime of the coasting electron beam was only of the order of milliseconds. Kerst provided components and the technology from his betatron laboratory for the electron injection. An electron gun (a filament within an anode shield) could be pulsed or operated dc at 25–30 kV. An orbit expander, produced by the discharge of a 0.05 μf capacitor by a pulsed thyratron, accelerated the electrons at about 400 V per turn for several turns, sufficient to miss the injector on their subsequent revolutions. The betatron core, made of 500 lb of laminated transformer iron, was driven by a 500 Hz rotary converter (scavenged from a Michigan X-ray laboratory), and provided about 40 V-per-turn accelerating voltage. Later, a dual-pulse system, similar to that used on the Illinois model, was employed for experiments on rf acceleration.

During this period, a new graduate student at Michigan, David Wilkinson, worked on the accelerator with Jones and Terwilliger. He had built a small electron cyclotron as an undergraduate, and was interested in accelerators. However, in those days, a doctoral thesis in accelerator science did not seem an option (especially as Jones and Terwilliger were still Instructors, not yet promoted to Assistant Professor). So he became a student of Dick Crane, under whom he earned his Ph.D. on the electron g-factor. His subsequent career is briefly described in the Sidebar on Students. Other subsequently successful graduate students who worked on the model included Mel Stewart and Rolph Scharenberg. Later on, Charles Pruett was hired as a postdoctoral scientist to work on the model, and remained a part of the MURA — and, later, PSL — staff until his retirement.

Purdue Magnet Construction

After the design had been fixed, the magnet construction and testing were carried out at Purdue, under the direction of Bob Haxby. The design required that the fields "scale," i.e. that both $k = (r/B)\ dB/dr$ and the magnet edge fringe fields have the same shape at all radii. To preserve scaling, the magnet apertures were angled to make the gap proportional to the radius, increasing from 4.0 to 6.75 cm. The field (from ~40 to ~150 G) was produced by distributed pole-face windings, glued with epoxy resin to aluminum plates, which were in turn bolted to the magnet iron. The magnets were fabricated at Purdue and delivered to Michigan, where they were completed, over the course of several months, in mid-late-1955.

Assembly (During the 1955 Summer Study) and Static Sigma Tests

During the 1955 Summer Study at Michigan, the first few magnets were installed over the completed vacuum tank. To get an early handle on the validity of the beta-tron oscillation dynamics, a pinhole injector was installed at the central azimuth of a radial-focusing magnet, a brass plate with several pinholes was placed at the

Charles Pruett (1928–)

Charles Pruett was an excellent experimental physicist, a fine colleague, a most pleasant personality, and a member of the MURA staff through the lifetime of the corporation. Later, at the University of Wisconsin Synchrotron Radiation Center, he became a renowned expert in ultraviolet beam transport and instrumentation.

Pruett obtained his undergraduate and graduate degrees from Indiana University, and was hired by Jones and Terwilliger in 1955 to assist in the construction of the Radial Sector Model at the University of Michigan. As Jones and Terwilliger were still teaching classes and had less than full time to work on the accelerator construction, Pruett's full-time involvement was very important and valuable. He was involved in the first operation of this model, including the studies of the beam stability as a function of the horizontal and vertical betatron oscillation tunes. When, in September 1956, the model was moved to the new MURA laboratory at Madison, Pruett moved there also, and continued the studies of orbit stability, betatron acceleration, and then of the rf acceleration and beam stacking.

Following the completion and publication of the extensive studies with the Radial Sector Model, in 1959, he joined the group working on the construction of the 50 MeV FFAG Model at the Stoughton site, including work on magnetic field measurements and the construction of magnetic field correction coils. After that accelerator became operational, he was involved in many studies of orbit dynamics, acceleration and beam stacking, stacked beam instabilities, wideband feedback damping of instabilities, beam extraction, etc. After that he worked on many aspects of the Tantalus storage ring.

In 1967, Pruett joined the staff of the University of Wisconsin Physical Sciences Laboratory (PSL), continuing to work on the Tantalus project. He later became involved in the UV and soft X-ray optics for the synchrotron radiation research with Tantalus, leading to his appointment as the Optics Group Leader of the Synchrotron Radiation Center, the group which constructed 12 monochromators and associated beam-lines. He retired in the early 1990s.

Pruett has a son and a daughter, both with Ph.D. degrees, and three grandchildren. He is married and enjoys photography, English country dancing, Scottish country dancing, international folk dancing, recreational vehicle travel, and camping.

corresponding azimuth of the next radial-focusing magnet (i.e., displaced by one sector), and a ruled phosphorescent screen was centered in a third radial-focusing magnet. The "rays" traced by the electrons (as deduced from the spacing of the images on the screen) enabled the calculation of the betatron tune, for vertical and horizontal oscillations. Later, with all magnets installed, the screen was installed one sector behind the injector, so that the rays were imaged after 7/8 of a revolution.

These ray-tracing measurements were also valuable for studying the dependence of the betatron tune with amplitude, the nonlinear oscillations.

The First Operation, March 1956

In early 1956, assembly of all components at Michigan was completed. With all systems functioning, an attempt was made to observe an accelerated beam. In fact, everything worked exactly as planned — and hoped — and indeed the accelerated beam was clearly observed (to the great thrill of Jones and Terwilliger). Kerst was telephoned, and he, too, rejoiced. The accelerated beam was typically about 10^8 electrons per microsecond pulse. Alternatively, the injector could be run dc at 25 kV, with a time average current of about 1.5×10^{-8} A, and a continuous beam pulse about 600 μs long every 2 milliseconds.

Studies of the Stability Diagram

The betatron acceleration took about 160 μs, although it could be slowed to a few hundred μs with reduced core excitation. Except for losses from gas scattering, the beam was quite stable. The beam could also be accelerated to an intermediate energy by pulsing the betatron core and later applying a pulse to accelerate it to an outer target. The design tune of the machine was for 2.80 radial and 1.80 vertical betatron oscillations per revolution, with $k = 3.36$. The machine tune could be varied by changing the ratio of currents in the positive-curvature (radially focusing) to the negative-curvature (radially defocusing) magnets from the design value of unity, and also by changing the ratio of currents through the coils on the back-legs of the magnets to that in the pole-face windings, which changed k (albeit in a non-scaling way). By this means, the tune could be varied through a band of about 1.8 units of vertical tune (oscillations per revolution) and 0.4 units of radial tune. The betatron oscillation frequencies were measured both with the static, ray-tracing measurements as described above, and with the dynamic "rf knockout" method, invented by Terwilliger. He had developed this technique for studying the betatron oscillations in Dick Crane's 70 MeV Michigan electron synchrotron. This involved applying an rf voltage to a probe near the circulating beam and varying the frequency until the beam was destroyed. This occurred when the applied rf was at a frequency corresponding to a beat frequency between the betatron oscillation frequency (vertical or horizontal) and the revolution frequency. These studies were mostly made with the beam close to the injector, coasting after being accelerated by the expander by a few kilovolts. By observing the intensity versus tune, a two-dimensional map was created showing the integral, half-integral, and coupling resonances in the horizontal and vertical betatron oscillations. These rf knockout measurements agreed with the static, ray-tracing measurements. However, at the design currents, the vertical oscillation tune was about 2.15, significantly above the design value of 1.80, although the horizontal tune was closer, about 2.86 (with a design value of 2.80). However, with the assembled magnets, the magnetic fields were mapped and, with them, the calculated tunes (using the later, more sophisticated computer codes) agreed quite well with the measured values.

The design of the FFAG with k constant required that the gradient, dB/dr, not be constant with the radius; hence nonlinearities were built in, and changes in betatron frequencies with amplitude were expected, and observed. The effects of misalignments were also studied. The magnets were positioned by precision steel dowel pins, which secured each magnet to the table. After removing the pins from a magnet, it could be moved radially or raised by measured amounts and the resulting perturbations on the accelerator orbit parameters measured.

The Move to Madison, September 1956

In the late summer of 1956, the Michigan Model was moved to Madison, where the MURA group had agreed to establish their center of operations. Jones and Terwilliger took a year's leave of absence from Michigan and joined the MURA group there, together with Pruett. Hence the work on the Michigan Model resumed there in the autumn of 1956.

The Addition of RF Acceleration; the Demonstration of Phase Displacement, etc.

At Madison, following the development of the Symon–Sessler rf theory, the betatron tune studies were completed and studies of rf acceleration, phase displacement, and beam stacking were undertaken. The revolution frequency of the beam, between injection energy (25 keV) and final energy (400 keV), was from 36 to 74 MHz. There was no space for insertion of an rf cavity in the machine, but rf could be applied across the insulating gap on the vacuum tank provided for betatron acceleration. Jones and Terwilliger had learned that Crane had used the ferroelectric properties of barium-titanate capacitors to make an electronically tuned rf accelerating system for the Michigan synchrotron, and they applied this same concept to the Radial Sector (Michigan) Model. With the vacuum tank as a one-turn inductor, and a set of $BaTiO_3$ capacitors biased by up to 3.5 kV, an rf oscillator was made, which could be tuned over a range of about 10 MHz, around 70 MHz. Studies were made by accelerating a beam bunch to about 300 keV via a betatron pulse, where it would coast for about 2 milliseconds. Over a fraction of a millisecond, rf was then applied to the beam, followed by a second betatron pulse to accelerate the beam to a target at the full-energy radius. The applied rf accelerating voltage could be up to 45 V. The time distribution of the beam at the target provided an analysis of the beam energy distribution following the rf manipulations (higher energy particles were the first to the target, so the arrival time was a measure of the beam energy). With this system, many studies were made of beam stacking, phase displacement, and other rf manipulations that were quantitatively correlated with predictions of the Symon–Sessler model. A good communications receiver proved to be a useful diagnostic tool, not only for calibrating the rf system and measuring the rf beam knockout frequencies (for determination of the betatron oscillation frequencies), but also for detecting various other beam-generated rf signals by locating a probe (antenna) within the vacuum tank.

Final Results

Jones and Terwilliger returned to Michigan during the autumn of 1957 and Pruett continued work on the Michigan Model at Madison. However the major results had been collected and published by then, and the experimental work at Madison turned to the Spiral Sector (Illinois) Model, and to the design of the 50 MeV Two-Way Model. As the first operating FFAG accelerator, and as a physical test bed for the orbit calculations, magnetic field designs, and rf theory, this small machine was a resounding success.

Keith Symon recalls being impressed that both the Radial Sector Model and the later Spiral Sector Model operated with beam when they were first switched on after assembly, without requiring any tuning or adjustments. This was probably a first for particle accelerators, and is presumably to be credited to the fact that they were designed carefully, with magnetic fields measured and compared with the design values, and orbits calculated before assembly.

3.7. THE SPIRAL SECTOR MODEL

Following the invention of this concept by Kerst in 1955, the MURA group decided to build another FFAG electron model [Kerst, 1960], this time of the spiral sector variety. It would use similar injection technology, injecting at about 35 keV, and its final energy was to be 180 keV, above the transition energy (the energy at which the revolution frequency is highest).

The Illinois Design Group

Kerst assembled a group including Laslett, Sessler, Snyder, and Peterson. A remarkable improvement in calculating magnetic fields was made for use on ILLIAC, the

1024-word digital computer at Urbana, Illinois. In addition, improved methods were made for evaluating the effects of construction errors. The requirements above, and careful consideration of the working point, led to a six-sector design with gradient index k about equal to 0.7, a spiral angle of about 46° with respect to a radius, and a flutter (rms field variation on the median plane) of about 1.1. These parameters led to tunes of 1.4 horizontally, 1.1 vertically, and transition kinetic energy of 155 keV. Since the circumference factor was only about 2, the peak fields were about three times lower than those in the Radial Sector Model. The lower field and the relative openness of the magnetic structure necessitated extreme care in avoiding and correcting for environmental magnetic fields. The accelerator as constructed is shown in Fig. 1.2. It was often referred to as the Illinois Model.

Component Construction

Magnet construction was started at Urbana and completed at Madison by Peterson and Haxby. The poles were cut from steel sheets forged into conical surfaces. The pole surfaces and current slots were fine-machined to the proper shape. The poles were then pinned and bolted to steel bars (back-legs) which carried magnetic flux between the top and bottom poles. The magnets were made up of spiral segments of conical steel sheets, with back-legs at large radius, and return current conductors at constant radius distributed along the pole to make the field along a spiral vary as the radius to the power k. The current in these windings was returned in slots near the edge of the pole to increase the flutter. The average field on the injection orbit was about 16 G. The gap at injection was about 9 cm. A large coil was made to cancel the environmental magnetic field in the region of the accelerator. The whole accelerator was mounted on a table with aluminum supports and a thick Masonite surface. Brass or stainless steel screws were used in all components except magnets.

The vacuum chamber and pumping system, built by E. A. Day and F. E. Mills, was made up of two hollow dees of brass fabricated by low-temperature brazing, and insulated from each other by rubber vacuum sealing gaskets compressed by insulated bolts. Supporting the relatively large span was already a problem (to become much worse in the later Two-Way Model), obviating the simpler aluminum solution in the Radial Sector Model. The pumping system was typical for its time, using oil diffusion pumps and yielding a pressure of about 3×10^{-6} Torr.

The plan at Illinois had been to braze the brass tank at high temperature with eutectic solder. When Ed Day attempted to do this in spite of advice to the contrary from several physicists, including Terwilliger, Jones and Mills, the result was a mess of distorted plates joined at several points, thoroughly unsuitable for its purpose. He then devised a scheme to use lower-temperature solder, followed by machining, which was a perfectly satisfactory solution, at least for this accelerator.

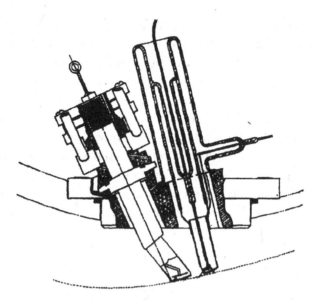

Fig. 3.16.   A drawing of the Spiral Sector Model injection system, showing the electron gun and septum.

The injection systems, developed by Peterson, Mills and Rowe, were considerably improved over that of the Radial Sector Model. A similar electron gun beamed the electrons through an electrostatic inflector, the whole assembly being mounted on a rotary seal so that the injection angle could be adjusted without breaking vacuum. The effective septum width was several tenths of a mm (Fig. 3.16). The injection voltage could be modulated by two different means. The first means was a thyratron pulsed delay line followed by a pulse transformer, similar to those in use at typical circular electron accelerators, giving a pulse length of several microseconds. The second system was a hard tube modulator with a blocking capacitor that allowed pulses of about 500 μs without appreciable voltage droop. This could test the idea of an FFAG betatron that might attain high intensity electron acceleration.

A shaped insulated deflecting plate, whose voltage was modulated by a pulse-forming network, was placed 180° from the injection septum to test multiturn injection.

The primary means of acceleration, built by Peterson and Mills, was a laminated iron "betatron" core, the secondary of which was the beam. The pulse length was much longer than the acceleration time, so the chamber could be filled with beam at all energies. The core was wound with distributed windings in parallel with "flux forcing" windings around the gaps in the iron core to reduce environmental fields from the system. (Failure to do this caused serious problems with the operation of the FFAG model at the Institute for Physical Problems in Moscow, USSR.) Energy stored in a capacitor was pulsed through the windings on the steel core, reversing the voltage on the capacitor core. A second current pulse reversed the voltage again, and the capacitor was recharged to its original value. In this way, the beam could be accelerated to an intermediate energy, acted upon by some other system, and the remainder accelerated to the target by the second pulse.

For rf studies, a system similar to that employed on the Radial Sector Model was developed and used by Peterson to study acceleration across transition energy.

A single-turn extraction system was built by Mills, Radmer, and Shea consisting of a fast kicker switched by a mercury switch (0.5 nsec rise time) followed by a focusing electrostatic deflector which brought the beam outward radially to an insulated fluorescent screen. To reduce acceleration drift during the jitter time of the mercury switch, a "tail biter" (a many-turn coil around the betatron core shorted by a thyratron) was fired at the end of the betatron pulse.

A timing, safety, and control system, built by Mills and Radmer, was used to sequence operation, and to prevent operation in failure modes.

The beam diagnostics were built by Mills and Day. A circular access port was located on the outside of the vacuum tank in each of the six straight sections, and in several of the inside ones. Insertable, rotatable, insulated probes were in most of these. They were simple flags to act as targets. Some had offset horizontal wires to measure vertical beam size. These probes could also be excited with rf signals to excite betatron oscillations in either horizontal or vertical motion. One straight section was devoted to a plastic scintillator glued into the end of a brass tube, with an aluminum coating evaporated onto it to keep out light. The scintillator was viewed by a photomultiplier producing a signal on a resistor shunted by a transmission line to an oscilloscope.

Operation and Research Program

Parts of the accelerator arrived at Madison from Illinois in the autumn and winter of 1956, after the laboratory facilities were in operation. The first accelerated beam was effortlessly observed in the autumn of 1957. Simply raising the injection voltage above threshold was sufficient to observe accelerated beam on the scintillator with only field emission from the cathode, i.e., without energizing the injector filament. Typically, the photomultiplier was operated at low voltage, and the injector filament turned up until a satisfactory operating current was obtained, while still having a proportional response from the photomultiplier. For higher currents, the beam current was intercepted on a flag. At higher currents one could observe saturation of the beam intensity due to space charge.

A map of intensity versus horizontal and vertical tune was made by Wallenmeyer that showed the effects of major resonances on operation. The working point was different from that of the Radial Sector Model, as seen in Fig. 3.15.

Carl Radmer (1934–1997)

Carl Radmer was an outstanding example of a young engineer who joined the MURA effort soon after graduation and then went on to a notable career in industry. MURA provided an opportunity for many young engineers to expand their horizons and develop their talents in state-of-the-art projects. Just as postdoctorial physicists spend the early years of their careers on research projects to broaden their interests and abilities, young engineers increase their skills by being exposed to a variety of research projects.

Carl joined MURA in 1957, after graduating from the University of Wisconsin with a degree in Electrical Engineering. He first worked on controls for the Spiral Sector Model. As time went on, he grew and grew as an engineer. He built extraction kickers and power supplies, spark gaps, power supplies for betatron accelerating cores, high-power radio frequency amplifiers for linacs, etc. Very soon after joining MURA, Carl was required to serve a term in the Army Signal Corps, but he returned to his work at MURA to make important contributions to the problem of injecting beam into the 50 MeV Two-Way Model. In addition, he participated in beam studies, designed beam measurement probes, and helped develop the control and protection system for the accelerator. When the emphasis in the laboratory turned to linac development, Carl designed the radio frequency power supplies and controls for the experimental program. Following the linac project, in 1965, he designed and built all the electronics for the Michigan–MURA cosmic ray experiment on Mount Evans in a period of six weeks, installed it on a 14,000 ft mountain and carried out the experiment. He was later responsible for the electronics of the Echo Lake cosmic ray experiment.

When Walter LeCroy, head of an electronics firm, saw what Carl had done, he offered Carl a job and just kept increasing the salary and company share until Carl said yes. In 1966, Carl joined the LeCroy Corporation as a vice president and continued there until his retirement in 1995. Sad for MURA, but great for the electronics industry!

In the corporate world Carl continued his interaction with scientists; his company produced electronic equipment for many of their experiments. He was a frequent visitor to all the leading accelerator laboratories and his advice was sought where large quantities of experimental data had to be rapidly recorded. Carlo Rubbia, an Italian physicist and the 1984 Nobel Prize winner, was reputed to have developed a high respect for the advice that Carl offered.

In retirement, Carl led an active life pursuing his many hobbies and traveling to many places around the world. He enjoyed his family and was a great husband to his wife, Ethel, and helped her pursue her notable career, as well as a devoted father to their three children, and later a doting grandfather. His untimely death was a severe blow to his family and his many friends.

The beam was accelerated across transition with the rf system in experiments by Peterson. The rf phase was shifted by raising the frequency slightly above the "synchronous" value while near transition energy. The results imply an efficiency of between 50% and 70% for the acceleration of captured beam through transition. The experiment was complicated by the relatively short (400 μs) lifetime due to gas scattering. The results are seen in Fig. 3.17. In all three cases, beam is injected and stacked by a previous betatron pulse. The intensity striking the target is shown in trace α, and the accelerating voltage, from either the betatron core or the rf system, is shown in trace β. In case (a) there is no rf and the beam is simply accelerated to the target during the second betatron pulse. In case (b), some beam is captured by the rf and stacked at the transition energy, and so arrives at the target before the uncaptured beam. In case (c), the captured beam is partially accelerated through transition and strikes the target before the betatron pulse. The originally uncaptured beam, and the beam not accelerated through transition, strike the target during the second betatron pulse.

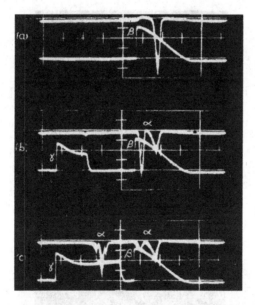

Fig. 3.17. Acceleration across the transition energy as observed in the Spiral Sector Model. In all three cases, beam is injected and stacked by a previous betatron pulse. The intensity striking the target is shown in trace α, and the accelerating voltage, either from the betatron core or the rf system is shown in trace β. In case a, there is no rf and the beam is simply accelerated to the target during the second betatron pulse. In case (b), some beam is captured by the rf and stacked at the transition energy, and so arrives at the target before the uncaptured beam. In case (c), the captured beam is partially accelerated through transition and strikes the target before the betatron pulse. The originally uncaptured beam, and the beam not accelerated through transition, strike the target during the second betatron pulse.

The long-pulse injection mode, developed by Rowe and Mills, worked as expected at low intensities and began to show interesting phenomena at higher intensities. For example, the tune could be set at a resonance — say, $v_z = 1$ — where at low intensity no beam was normally accelerated. As the beam was neutralized by ionization, it would be accelerated, thus demonstrating the ion-induced tune shift. Operating normally, as the injected intensity was increased, the whole radial aperture would be void of beam in a period of a few microseconds. Subsequently, the beam would be accelerated again and, after a while, voided again. This happened repeatedly, depending on the intensity and total pulse length, as can be seen in Fig. 3.18. This phenomenon was studied by Don Roiseland as part of his Ph.D. thesis but was never fully understood in the context of the instability theories developed later.

The single-turn extraction system developed by Mills, Radmer and Shea worked well, and yielded a focused beam at the screen. It was also observed that if the kicker

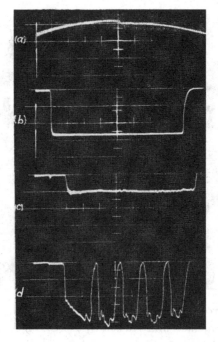

Fig. 3.18. A space charge instability as seen in the Spiral Sector Model. Trace (a) is the betatron accelerating voltage, trace (b) is the injector voltage, trace (c) is the beam intensity striking a flag when the intensity is low, while trace (d) is the intensity striking the flag when the intensity is high. Several phenomena are evident. The initial intensity is limited by space charge, and cannot be increased by raising the injector current. Subsequently the intensity increases as ions are trapped in the beam. Finally, the repetitive instability phenomenon starts. During the beam loss, rf signals could be seen, but they appeared to be at the revolution frequency of some energy in the aperture as in Fig. 3.17.

was not fired and the beam accelerated slowly, part of the beam was extracted by scattering on the septum.

3.8. MURA COMPUTING

The MAC and MURA physicists began to use digital computing very early in the activity. The University of Illinois was an early leader in computing, having built ORDVAC for the US Army Ordnance Laboratory, followed by ILLIAC, a slightly improved version of ORDVAC. ILLIAC had a memory of 1024 40-bit words using Williamson cathode ray memory tubes, numerous registers, and an extended order code. An 18-word Decimal Order Input routine converted decimal numbers to binary or floating-point numbers as needed. Input and output were accomplished by punched paper tape. Until 1956, most MURA computing was done using this computer. The programs were mainly of three types: relaxation solvers for magnetic potentials and fields, integrating programs using the Runge–Kutta method to solve nonlinear differential equations, and programs to simulate acceleration schemes by using successive transformations. There were also algebraic transformation programs to investigate nonlinear phenomena.

When the laboratory at Madison started operation, an IBM 704 was rented for use by the laboratory. The 704 had high-speed magnetic core memory. In addition it had magnetic core memory for storing parts of a program and tapes for bulk storage (and for input). Input was by magnetic tape/punched card/console. Output was mainly a line printer or punched cards. The word length was 35 bits plus sign. The speed for a multiply/divide was 240 μs, or 4000 operations per second. Input and output were via punched cards. The ILLIAC programs were converted and improved. Eventually, relaxation programs to calculate fields, losses, resonance frequencies, and particle motion in rf cavities were added. As a then novel application of computers, S.C. Snowdon in 1963 used the results of the magnetic relaxation programs to compute the path of a milling machine head to produce the surface of a magnetic pole that produced the proper field for an FFAG accelerator. The magnets were constructed and tested as part of the modeling program for the 12.5 GeV accelerator. (The MURA proposal had been for 10 GeV, upgraded to 12.5 GeV in 1962 at the request of the AEC.)

When the 704 was first rented, AEC had a policy that computers could not be purchased with AEC laboratory funds, but only rented. The MURA management found that the computer could be purchased with rent of a reasonably short period, and worked out a plan with AEC for the MURA Corporation to purchase the computer and rent it to AEC at the same rate until the computer was paid for, at which time the rate for that use became nominal. However, the computer was better than any available to research labs, industries, and many university grants,

Fig. 3.19. A 10 GeV spiral sector development model magnet used for magnetic measurements in 1962. This design allowed the introduction of radial straight sections into the spiral sector geometry.

and the MURA Corporation rented time to other users when the computer was not in use for AEC purposes. Eventually, AEC decided that, in spite of the fact that this had been done *because* of AEC rules, *AEC* should receive the income from the rental. This caused many complications for the Corporation, delaying its disbandment after the laboratory was no longer supported by AEC. The Corporation used the money in a way consistent with the initial assessment paid by each university when the Corporation was formed to assist in the development of the laboratory. In addition to the purchase of the land on which the laboratory was located, and the 20,000 ft² office and laboratory building, one of the uses (US$6,000) was to erect the temporary housing where the 240 MeV storage ring was assembled and operated.

Following is a recollection by A. M. Sessler of computing by MURA members at Urbana:

"The MURA Group had sole use of the ILLIAC one evening a week. The ILLIAC received information from paper tape prepared by a Teletype machine. The tape was fed in by hand and read with an electronic eye. Similarly, the output was punched paper tape, which we would quickly run over to a Teletype machine (there were a number in the room), feed in the tape and see the printed version.

Stanley C. Snowdon (1918–1992)

Stanley C. Snowdon was the magnet designer *par excellence* who developed the means of designing and building the magnets for the large FFAG accelerators described in the MURA proposals. Working with Dick Christian, he developed the concept of "integral scaling," which allowed the construction of spiral sector accelerators with radial straight sections. He went on to design many types of magnets at Fermilab, of which thousands were built and operated. His work was widely copied.

Stanley was born in Arlington, Massachusetts, of Scottish descent. He was awarded a full scholarship to study at the Massachusetts Institute of Technology (MIT) in 1936. Upon graduation in 1940 from MIT, he was given a teaching fellowship at the California Institute of Technology. He obtained his Ph.D. degree with *magna cum laude* honors in 1943. He was employed as a Research Associate at Caltech and the University of Wisconsin, at the Bartol Research Foundation as a staff member and concurrently as a lecturer at Temple University, before joining MURA in 1959. He served as a consultant for various organizations on a variety of projects. He had over 100 publications. He was a Fellow of the American Physical Society.

With the invention of the FFAG accelerator came more rigid requirements on the specifications and tolerances of the accelerator magnetic fields. These problems were addressed with the growing versatility of digital computers and computer programs to calculate field shaping in iron-dominated and conductor-dominated magnets. Stan arrived at MURA in 1959 to address these problems. A number of MURA reports were written describing his efforts on magnet design. In 1967, at the beginning of the National Accelerator Laboratory (now Fermilab), Stan gave up an opportunity to become a professor at the University of Wisconsin, to continue the work started at MURA on magnetic field calculations and development, thereby providing a necessary and valuable role in the design of the many magnets required in the Fermilab accelerator complex.

Stan led an active life with his family and church. He and his wife, Betty, had three children, and were active in camping and scouting, having been in scouting for over 19 years. His latest hobby was the reassembly of a sport automobile after a complete rebuilding.

The ILLIAC had electrostatic storage, so there were a good number of dark TV-like tubes around the top of the machine where storage was done.

On the evenings when we had use of the ILLIAC, usually about a half dozen of us would gather in the room and use the ILLIAC to help our 'thinking'; that is, the machine was not used for detailed evaluation, but rather to answer qualitative questions. Thus there was real interaction with the machine. We would try out something, and then on the basis of the output try something else. It was really using the computer to help in a creative process.

The rest of the week we used the ILLIAC in a batch mode in which we did more what nowadays would be called number crunching. We had a few programs at that time and we did calculations overnight, typically only once a day, although sometimes we could put in information in the early morning and get the results in the late afternoon. The input was organized in single printed pages in which we entered the desired parameters of the run. A young person, Jesse Anderson, was the 'gofer'; he would pick up the sheets, transfer the information, by Teletype machines, into paper tape and then do the reverse with the output. He would pick up the sheets in our office and bring back the output to our office.

We became quite friendly with Jesse and he helped me build up a record collection of fine classical music and now, some 50 years later, I still listen to, and enjoy, the records he selected for me. It was a wonderful extension of my classical music appreciation, for which I am very much in debt to him."

Another recollection, by Frank Cole, now deceased, was related to Fred Mills:

"When we first started integrating orbits through repeating structures, we expected to see values of position and angle at corresponding points in the cell plot out ellipses with a fixed orientation. Much to our surprise, the ellipses 'rolled,' i.e. their orientation changed with time. Quickly we discovered the problem: the value of pi used in the program was too approximate. When this was corrected, the ellipses became stationary. We decided that Don [Kerst] had gotten the original value by reading it off his slide rule."

Fred Mills also has a recollection of ILLIAC computing of that era:

"By the summer of 1954, I had taken the data for my thesis using the University of Illinois 300 MeV Betatron, and was analyzing the data. I decided that the best way to estimate the efficiency of the detector was by the 'Monte Carlo' method following the sequence of events leading from an interaction to a successful detection. The result of each step was chosen at random from the appropriate probability distribution for that step.

At that time a loudspeaker was connected to the sign bit in one of the computer registers, and each program had its own sound. The operator could follow the progress of the program and stop it if it sounded wrong. When the operators put in my program, there was no pattern, only a mumbling sound, so they immediately switched it off. Eventually I convinced the operators that the program sounded just as it should, and I completed the calculation, much to the relief of the operators."

3.9. COLLIDING BEAMS

It was known, even prior to WWII, that at very high energies, colliding energetic beams would provide a much greater available energy than an energetic beam impinging on a fixed target. In a collision, the energy available to produce a reaction is the center-of-mass energy, defined as the total energy in the center of mass coordinate system (the system where the total momentum is zero). For a relativistic particle of kinetic energy E_0 colliding with a stationary target of mass m, this energy is given approximately by

$$E_{cm} = (2mc^2 E_0)^{1/2}, \qquad (3.5)$$

where we assume the rest energy mc^2 is much less than the kinetic energy E_0 supplied by the accelerator. This increases only as the square root of the kinetic energy, so to double the center-of-mass energy, we would have to quadruple the accelerator energy. In a colliding beam experiment, where two identical particles have equal and opposite momenta, the experiment is carried out in the center of mass, so the center-of-mass energy is $2(E_0 + mc^2)$, or approximately $2E_0$. To double the center-of-mass energy, we need only double the accelerator energy.

Kjell Johnsen, a pioneering accelerator physicist at CERN, noted — many years later — that this had been assigned as a homework problem when he was in graduate school. But the idea of colliding two accelerated particle beams to achieve high-energy interactions was not seriously considered. During WWII, the concept of colliding beams was, in fact, patented by Rolf Wideröe in 1943 (patenting was the only way available to publish scientific advances in wartime Nazi Germany). However, despite these thoughts, no one had the slightest idea of how to make the beams sufficiently intense to provide a practical, interesting collision rate between them (note the reference to Wideröe at the 1956 CERN Conference in Sec. 3.11). And, before the construction of the Brookhaven AGS and the CERN PS (completed about 1960), the proton–proton interaction energies were not so highly relativistic that beam–beam collisions would offer a significant advantage. For example, the Bevatron kinetic energy of ~ 6 GeV was chosen to be above the 1.88 GeV threshold (available c.m. energy) required to produce a proton–antiproton pair.

During the autumn of 1955, the MURA Working Group was in residence at the University of Illinois (Sec. 3.3). It was early in this period that Kerst first considered seriously the idea of colliding beams, in view of the potentialities of FFAG machines to achieve high-intensity circulating beams. Jones recalls an hour-long phone call from Kerst to him and Terwilliger at Michigan in early September, where he discussed the colliding beam concept, its promise, and the related problems. Some of these ideas were significantly clarified by Kerst's visit to Princeton, where he presented a colloquium, and where Wigner discussed with him the importance of phase space and of Liouville's theorem in the analysis of beam stacking, as described in Sec. 3.6. These ideas were discussed intensively by the Illinois Working Group. Sessler recalls that, when he joined this group during that autumn of 1955, the very first questions that Kerst asked him were: (1) How does one manipulate the rf so as to build up an intense beam without destroying the stacked beam at high energy? (2) Will nonlinear behavior allow the stacked beam to last for a very long time? (See Sec. 3.5.)

With some assurance of long-term stability and with some understanding of stacking, MURA could now, for the first time, seriously propose a proton–proton colliding beam accelerator system. At the MURA general meeting in early October, 1955 at Urbana, which attracted about 70 high-energy and accelerator physicists from all parts of the US, Kerst presented his ideas. Later, after the studies had verified the real possibility of achieving a practical collision rate, he wrote a Letter to the Editors of the *Physical Review* [Kerst, 1955]. Although the basic idea and its practical realization were really the work of Kerst himself, he very generously included much of the MURA group as coauthors. The basic geometrical configuration, as in the illustration in this Letter, was two FFAG accelerators tangent to each other, sharing a common straight section in which the beams from the two accelerators (traveling in opposite directions) would collide.

Tihiro Ohkawa at MURA proposed a radial sector accelerator in which the two magnets in each sector are identical except for having fields in opposite directions [Ohkawa, 1958]. He pointed out that the orbits would be closed because the orbit is farther out (higher field) in the positive magnet. However, in view of the symmetry, particles can now circulate in either direction. This two-way configuration was clearly of interest for colliding beams.

Of course, this concept was widely discussed during that autumn. Keith Symon recalls being invited to give a colloquium on this subject at the University of Illinois, and when he first mentioned colliding beams, the audience burst out laughing. He was somewhat taken aback, but later learned that at the colloquium the previous week, Kerst and Gerry Kruger had illustrated the colliding beam concept by shooting at each other from opposite sides of the stage with peashooters.

At a MURA general meeting at Indiana University in February 1956, three of the university's physicists, not closely connected with MURA, proposed the use of storage rings for colliding beams, such as transferring the accelerated beam from the FFAG accelerator to a pair of tangent storage rings, in the common straight section of which the beam–beam collisions would occur [Lichtenberg, 1956]. Up to this time, the extraction of full-energy beams from accelerators was rather primitive; the only high-energy accelerators then operating in the US were the Brookhaven

Cosmotron and the Berkeley Bevatron. To extract the proton beams, they utilized the "Piccioni extraction" method (suggested by Oreste Piccioni and, independently by B. T. Wright), where the beams were accelerated into an energy loss absorber at the outer radius of the vacuum tank, following which the reduced-energy particles' orbit put them through a magnet (at the inner radius) which deflected them into an external beam channel [Piccioni, 1955; Wright, 1954]. This was quite inefficient; the protons were scattered and many interacted in the energy loss absorber, so that the intensity and phase space density of the extracted beam were significantly poorer than in the accelerator. The vacuum tank aperture was so large and the time of revolution so short that (in those days) it was impractical to pulse an electromagnetic system to cleanly extract the beam in one turn. Probably for these reasons, the MURA group in 1956 remained more interested in the tangent FFAG accelerators as means of achieving colliding beams.

At Princeton, Gerard O'Neill independently conceived the storage ring idea, which he published that spring [O'Neill, 1956]. He proposed stacking the accelerated beam in a storage ring by using an array of tapered wedges within the storage ring aperture; the injected beam would lose energy in the wedges, rapidly at first (where the wedges were thicker, at the larger radius), and then at a smaller radius, near the center of the magnet aperture, the wedges tapered to zero thickness; the idea was that the beams would accumulate at that radius and energy. Although Symon later showed that this scheme violated Liouville's theorem, this was the scheme that O'Neill presented at the 1956 CERN Accelerator Conference (as noted in Sec. 3.11). Another idea of O'Neill was, in fact, much more practical, and was an example of the concepts which would replace the Piccioni extraction concept on the larger, strong focusing machines with their smaller aperture. He suggested constructing a deflecting magnet as an element in a lumped transmission line. This method led to wave forms with faster rise and fall times, permitting transfer of the whole beam from one ring to another. In the US, it was very probably Matt Sands' discussion of transferring beams from lower-energy to higher-energy accelerator rings in 1959 (Sec. 4.6) which made the storage ring concept the more generally accepted means of achieving colliding proton — or proton and antiproton — beams. It would develop somewhat later that the ability to incorporate special straight sections into AG structures would be a decisive advantage for the AG storage rings. Later, it was learned that Toshio Kitagaki had earlier considered cascading synchrotrons. In fact the first system which used one ring to charge another ring was the 100 MeV electron–electron collider at the Institute for Nuclear Physics (INP) in Novosibirsk. The second was the PSL storage ring charged by the MURA 50 MeV FFAG.

The MURA group remained focused on proton accelerators, and on proton–proton colliding beam machines; however, the advantages of colliding beams were obvious for the case of electron–electron and electron–positron collisions.

First, because of the small rest mass of the electron (~ 0.5 MeV/c^2) and because electron collisions with matter are complicated by the nuclear EM interactions, beam–beam collisions of electrons are attractive at energies above only a few MeV per beam. Further, because of radiation damping, the phase space density is not conserved, and emittance preservation is not as critical as with proton machines. Thus O'Neill very early interacted with the HEPL (High Energy Physics Laboratory, at Stanford) community to build an electron–electron collider there. They built a pair of tangent rings of 500 MeV each (500×500), completed in 1963. Gersh Budker pursued a vigorous program at INP, where his group moved and operated the VEP1 collider, completed in 1963. And Bruno Touschek in Italy actually built the first operating electron–positron colliding beam machine, ADA, at INFN Laboratory in Frascati, colliding 250 MeV electron and positron beams; it first operated in 1961 [Barber, 1966; Budker, 1966]. Of course, the MURA 50 MeV two-way (Mark Ia) electron model was designed to enable electron–electron collisions, although it was never operated as a colliding beam accelerator. At CERN, to better understand the physics of storage rings, an electron model of the Intersecting Storage Rings (ISR) was initiated in 1960. Although it could operate at 100 MeV, it was usually run at about 2 MeV. This energy was low enough for radiation damping to be negligible, and therefore the ring was a good model of proton behavior. One was thus able to check — to some degree — the long-term stability of the beam [Ferger, 1963].

With the successful operation of the electron model, CERN proceeded with its high-energy proton–proton collider, the ISR. This took full-energy (28 GeV) beams from the CERN proton synchrotron (PS) and injected them into the two concentric intersecting rings that crossed at eight points, providing up to eight potential experimental areas. The ISR was an adventurous machine to build; after all, single-particle stability might not be as was thought (it had only been minimally studied both theoretically and experimentally), and various other bad effects might occur. The machine, thanks to Kjell Johnsen's insistence, was conservatively built in all conventional regards. Thus one could focus on the new phenomena, and if necessary, the machine had the extra capability to handle any untoward effects. And there was one, namely an unexpected dependence on gas pressure, but, because of the conservative design, the vacuum could be improved by two orders of magnitude over the design value, and the ISR performed as predicted — in fact, eventually, even better than predicted commencing operation for experiments in early 1971 [Johnsen, 1971].

Sessler had been at CERN in 1966–1967, and was an active participant in the ISR program. Symon spent the 1962–1963 academic year at CERN. Jones was also at CERN for two years; 1961–1962 and 1965, plus shorter visits, primarily involved in strong-interaction spark chamber experiments at the 28 GeV proton synchrotron. During these visits he engaged in the ISR activities, including discussions regarding the experimental utilization of colliding beams [Jones, 1963B]. He recalls discussions with the then director, Victor Weisskopf, who enthusiastically supported the concept of colliding beams. In fact, Weisskopf had strongly backed the controversial decision that the next major accelerator at CERN should be a colliding beam machine rather than a fixed target accelerator [Kaiser, 2007].

The interaction rate at a colliding beam accelerator or storage ring system, expressed as the number of interactions per second, can be expressed as a product of the beam currents in the two intersecting beams divided by their cross-sectional area (and other factors), multiplied by the interaction cross section [PDG2006]. Of course, this expression, in a form appropriate for bunched beam lepton colliders or continuous beam hadron colliders, was included in the early papers by Kerst, O'Neill, and other pioneers. However, in the late 1960s, the term "luminosity" was coined, and L, the luminosity, was defined as the combination of machine parameters which, when multiplied by the cross section, defined the interaction rate. Hence, the interaction rate, R, is given by $R = L\sigma$. Currently, luminosity is the universally recognized designation for the capability of a given collider. The time-integrated luminosity represents the productivity of a collider over a given period of time, often expressed in terms of "inverse picobarns" (pb^{-1}) or "inverse femtobarns" (fb^{-1}). The concept of luminosity is sometimes used for fixed target experiments as well.

To increase the luminosity, it is necessary to decrease the beam size (cross-sectional area) in the interaction region, i.e., to have the betatron phase space occupied by the beam as small as possible and the dispersion small or zero. In AG lattices, it is possible to manipulate the beam size in both the vertical and horizontal directions by the use of special insertions. In FFAG accelerators this was never possible, primarily due to the nonlinearities in the magnetic fields. This gave a signal advantage to the AG colliders.

Proton–proton colliding beams were certainly a topic of discussion in the US in the years around 1960. For example, the Ramsey Panel (Sec. 5.8) in January 1963 invited Jones and O'Neill to discuss with them colliding beams, the technical feasibility, and the physics research potential. Jones and O'Neill met before the meeting and gave a strong presentation in support of proton storage rings. Their presentation went very well; however, the Panel did not, in their final report, recommend proceeding with any program on proton colliding beams.

In summary, although MURA never was able to build a colliding beam machine, it was instrumental in the construction of the CERN ISR, which then emboldened people to build antiproton–proton colliders, first at CERN (the Spp̄S) and later at Fermilab (the Tevatron Collider). The ill-fated Superconducting Super Collider (SSC) was to have been a 20-TeV-on-20-TeV proton–proton collider; the Relativistic Heavy Ion Collider (RHIC) at Brookhaven is a proton–proton and nucleus–nucleus collider in operation since 2001, and the CERN p–p Large Hadron Collider (LHC) is scheduled to commence operation in 2009. And at DESY (Germany), the Hadron Electron Ring Accelerator (HERA) ring enabled studies of electron–proton beam–beam collisions. Thus Kerst, Budker, Bruno Touschek, Kjell Johnsen, and Gerry O'Neill were the individuals and the motivating forces that brought about the development of colliders. The laboratories that supported this speculative (at that time) activity were MURA, Stanford, the Cambridge Electron Accelerator, Orsay, Frascati, CERN, and Novosibirsk. Kerst and the rest of the MURA group were instrumental in the development of proton–proton colliders, while Budker, Touschek, and O'Neill, with their respective groups, spearheaded the development of electron–electron and electron–positron colliders. Most of this happened during the decade between 1955 and 1966. Prior to that time, there were no colliders, while by 1965, a number of small devices had worked, and one could speculate, as Budker did, that in "a few years, 20% of high energy physics would come from storage rings" (a gross underestimate); now the majority of frontier high-energy physics (except for neutrino physics) is done with colliding beams. For example, while the discovery of the Ψ and J particles was made simultaneously at Brookhaven in fixed target mode and at the SLAC SPEAR e^+–e^- storage rings, the discovery of the W and Z bosons was made only at the CERN Sp$\bar{p}$S collider; these discoveries were rewarded with Nobel Prizes for four people.

A more extensive discussion on the history of colliders is available in the book by Pellegrini and Sessler, which contains copies of many of the relevant publications plus an interesting collection of photographs [Pellegrini, 1995]. The "Review of Particle Physics" [PDG2006] has a section on "Accelerator Physics of Colliders," containing a three-page table of "High Energy Collider Parameters," including all present and recent e^+ e^- and hadron colliders.

3.10. COLLECTIVE INSTABILITIES

Particle beams exhibit what are called "instabilities," or growing collective motion of the particles in the beam. This phenomenon usually involves interaction of the charges and currents in the beam with the medium surrounding the beam, creating forces on the beam itself that support and enlarge the collective motion. One recognizes the similarity of this process to that in which a column of soldiers, marching in step with one another, can excite dangerous or damaging motion in structures such as bridges. The solution, for the soldiers, is to march out of step so that the effect of one soldier cannot reinforce that of others. For beams, the solution is the same — to cause the particles to oscillate at different frequencies. This is called "Landau damping" in honor of Lev Landau, who looked at similar processes in plasmas. The growth of the instability is usually prompted by the presence of dissipation in the medium, which causes a time lag in the excited fields, which can accomplish the transfer of energy from (the larger source) longitudinal motion to transverse motion, or the redistribution of energy in the longitudinal motion.

At MURA, several important beam instabilities were encountered. The first, predicted by Nielsen, Sessler, and Symon [Nielsen, 1959A], was in the longitudinal motion. The prediction came about after a number of years spent trying to understand a longitudinal space charge anomaly. Carl Nielsen and Andrew Sessler shared an office at Ohio State University and they started in 1958 to work on the longitudinal space charge effect upon particles trapped in an rf bucket. This work was a natural continuation of the previously developed rf theory and, because it was one-dimensional, was rather easy to calculate and understand. They found that the methods they developed worked just fine below transition energy, but not above. In fact, they could not obtain an equilibrium situation above transition. Sessler argued that they should press ahead and publish the work, since it was well known that space charge effects were important only at low energies, so that the fact that the method broke down above transition was not of practical interest. The work was, in fact, published [Nielsen, 1959B].

Nielsen kept worrying about the situation above transition and, in order to consider a simple case, considered a uniform coasting (no rf) beam, but with regard for the effect of space charge. Using a hydrodynamic model, he soon convinced himself — and then others at MURA — that the beam would be unstable against bunching (no matter how tenuous the beam was). Sessler found this result unbelievable and put himself to the task of developing a better analysis than hydrodynamics afforded. He realized that the methods of plasma physics (just being developed at that time) were relevant, but he did not know any plasma physics. He spent three months doing no research, but learning the methods employed in plasma physics to study instabilities (which included the concept of Landau damping). An analysis was produced over the Christmas holidays of 1958. Then Keith Symon took over and did a complete and very elegant analysis, which was presented, with the three as coauthors, at the CERN Accelerator Conference in June 1959 [Nielsen, 1959A]. A very similar analysis was independently derived by A. Lebedev and A. Kolomenski and presented at the same conference [Kolomenskij, 1959].

The instability can be described very easily. It happens at energies higher than the "transition energy," where particles at higher energy proceed more slowly around the circumference than those at lower energy. In this case, the *repulsive* electrostatic force between particles causes them to bunch together rather than disperse. The instability is not easily controlled and causes the beam energy spread to increase. This limited the amount of beam that could be accumulated in the MURA 50 MeV accelerator.

Subsequently, the nonlinear behavior of the negative mass instability was observed and then studied by Mark Barton and Carl Nielsen [Barton, 1961]. Barton was working on the Cosmotron and his observations of nonlinear clumping of particles

Carl E. Nielsen (1915–2005)

Carl E. Nielsen worked primarily on theoretical aspects of the MURA program from afar; that is, while remaining as a professor at Ohio State University (OSU). In fact this was the pattern of a number of the professors who were associated with MURA in its early days with no permanent staff. They remained at their universities, worked on MURA things, partook in meetings, and often spent vacation periods from the universities, and the summers, at MURA. Nielsen, in fact, spent a year at MURA (1960–1961), but his best work was done while he was at OSU.

Nielsen made a number of contributions to MURA, but his most significant contribution was the realization — and initial hydrodynamic estimate — that beams of particles could become unstable at a level of intensity far below that required for equilibrium instability. No one had even dreamed of that possibility before Nielsen's work. As described in the text, he later teamed up with others who showed that Landau damping provided a threshold below which there was no instability, whereas far above threshold the hydrodynamic growth rate was valid.

Nielsen was born in Los Angeles, California. He received an A.B. in 1934, and a Ph.D. in 1941 from the University of California at Berkeley. His Ph.D. was obtained as a student of Robert Brode in cosmic rays, but during his graduate student days he learned a great deal about gaseous discharge, which nowadays is a branch of plasma physics. It was this knowledge that led him to appreciate that particle beams could be unstable. He went to OSU in 1947 after teaching at the University of California from 1941 to 1946. (During World War II, he was a conscientious objector.) He spent the academic year 1946–1947 as an Assistant Professor at the University of Denver.

His research interests were very broad, ranging from cosmic rays and cloud chambers to the physics of fluids and plasmas. He was also a dedicated teacher and even established, by gifts to the university, an Undergraduate Physics Research Endowment Fund, which provides scholarship support for undergraduates in Physics at OSU.

After MURA, Nielsen developed an interest in alternative energy sources and did much of the pioneering work on salinity-gradient solar ponds. He built and operated the longest-running demonstration solar pond in the US. Besides making many international consulting trips, he was a coauthor of a book on solar ponds. His work was honored with the Charles Abbot Award of the American Solar Energy Society.

He enjoyed partaking in the Agency for International Development (AID) Summer Institutes in India (1962, 1965, and 1968). He was a consultant to ORNL, ANL, TVA, and LLNL and spent a year at CERN, half a year at Culham and four summers at the Max Planck Institute in Munich. He was the author of more than 80 scientific papers.

(Continued)

Carl E. Nielsen (*Continued*)

Nielsen was raised in Paradise and Sacramento in California, and as a teenager, he worked on his family's chicken and apple farm. Both of his parents were teachers and becoming educated was very much the environment at home. On moving to Ohio, Nielsen and his wife of 67 years, Imogene (Jean to all), bought a large, wooded area outside Columbus and built a "California" style home, with natural wood and large areas of glass, that was heated with a heat pump operating from a pond on the property. They kept warm and raised three children in this home.

and clumps passing through each other, without disturbing each other, were far beyond the theory (which was only linear) and are a subject still of interest today.

At a meeting in Madison, Nielsen's observation of the bunching of protons in the Cosmotron due to their mutual Coulomb repulsion had been presented and discussed. On returning to Michigan, Jones and Terwilliger related this super-ficially contradictory result to their luncheon table colleagues, which on that day included George Uhlenbeck. Uhlenbeck shook his long finger at them and said, "You should read Maxwell's prize paper on the rings of Saturn, where exactly the same phenomenon — albeit with the opposite sign — occurs." Indeed, the mutual gravitational attraction between the particles in Saturn's rings separates them, as the trailing particle is accelerated, hence moving to an orbit with a larger radius and a longer period, while the leading particle is retarded, putting it into a smaller orbit with a shorter period, hence moving ahead. Maxwell had written this correct explanation of the diffuse nature of the rings of Saturn in 1838, in a paper that had won the Adams Prize [Maxwell, 1856] early in his career, before his studies of electromagnetism. Jones and Terwilliger communicated this discussion and analogy to their MURA colleagues, and indeed, henceforth the particle bunching in an accelerator ring due to Coulomb repulsion was referred to as the "negative mass instability." In the paper on "Longitudinal Instabilities in Intense Relativistic Beams" presented by Symon at the 1959 CERN Accelerator Conference [Nielsen, 1959a], Maxwell's paper is indeed appropriately referenced.

Following the MURA work, a very extensive calculation was carried out by Kelvin Neil, as his thesis, studying the interaction of an intense beam with an rf cavity [Neil, 1961; Laslett, 1961]. This work was performed at Berkeley, but very much had the involvement of former MURA members (who had moved there).

The MURA group soon turned its attention to transverse instabilities, as did the group at the Stanford electron–electron storage rings. The former discovered the instability in an unbunched beam [Pruett, 1963], verified the modes and the effect of Landau damping [Swenson, 1963], and inferred from the slow growth rate that the instability growth was provoked by the resistivity of the vacuum chamber walls [Curtis, 1963].

In fact, Nielsen and Sessler were stimulated by close interaction with the Stanford electron–electron colliding beam machine, where a slow-growing transverse instability had been observed in a tightly bunched beam. However, they worked very slowly. There was no rush in those days, and since both had "daytime jobs," the study was done, strictly for the fun of it, in the evenings, at one home or another. Usually some beer was at hand, so the work took place for only a short time. Nevertheless, after eliminating one effect after another, and out of desperation, they came to the idea of wall resistivity as the cause of the instability [Niel, 1965; Laslett, 1965].

The instability was in the transverse direction at frequencies related to the transverse oscillation frequencies. Nodes, or loops (sinusoidal variation of transverse beam position in integral numbers), grew with time. The growing modes were those where the number of loops was greater than the number of transverse oscillations per turn. (This allowed energy transfer from longitudinal to transverse.) In addition to Landau damping, this instability can be controlled by detecting the beam position and providing a deflecting field at a different place at the time the measured beam passes it. This is called "beam feedback" and was an important part of achieving the 10 A beam current in the 50 MeV ring in 1963 [Pruett, 1965]. Shortly thereafter, MURA staff members, together with ANL staff members, used this technique to cure the same instability in the ZGS [Martin, 1965]. This instability, as was the negative mass, was seen in the single-bunch mode also in the Cosmotron (Carl Nielsen, private communication). Related collective effects were also observed, such as tune shifts due to beam charge and currents, as well as trapped ions. These are described in more detail in sections about specific model accelerators.

Subsequently, Sessler and Courant [Sessler, 1966A] developed the theory for a bunched (rather than continuous) beam. Further work, by Vaccaro and Sessler [Sessler, 1967], showed how the concept of *impedance* could be introduced, and this concept is now in wide use.

3.11. CONFERENCES

The physics community has always very strongly valued discussions — correspondence, meetings, and electronic communications. During the Cold War between the West and the Soviet Union, there was a remarkably close interaction between American and Soviet high-energy and accelerator physicists. The fact that the World Wide Web arose to serve the needs of physicists also demonstrates this communications affinity. Another example is the formation of CERN, the close partnership of only recently conflicting nations in a trans-European research collaboration.

Of course, the MURA physicists were no exception, and it is of relevant interest to note some of these national and international meetings and interactions. The first significant product of the MURA group was the invention of the FFAG concept, and this was first presented to the broader American physics community at the (then annual) New York meeting of the American Physical Society in January 1955. Members of the Michigan working group (which had been meeting during the fall of 1954) presented four ten-minute talks on the FFAG concept — by Symon [Symon, 1955A], Jones [Jones, 1955A], Terwilliger [Terwilliger, 1955A], and Kerst [Kerst, 1955B]. Although these were only short talks in one of the many parallel sessions, the session was very well attended, and the talks were followed by numerous lively questions and discussions. Later that year, at the Thanksgiving meeting of the American Physical Society in Chicago, five more ten-minute papers were presented — three dealing with the Michigan Model design and early beam diagnostics, and the others by Symon and Ohkawa [Terwilliger, 1955C; Cole, 1955; Jones, 1955B; Symon, 1955B; Ohkawa, 1955].

In March 1956, Jones and Terwilliger had achieved the first accelerated beam in the Michigan Model, and had telephoned Kerst about their success. A week or so later, they had a phone call from Kerst, saying that one of them would be able to join him and Symon in attending an international conference on accelerators in Geneva, Switzerland, under the auspices of the recently organized CERN. They tossed a coin to choose who would go, and Jones won. This Symposium on High Energy Accelerators and Pion Physics was held in June 1956, on the campus of the University of Geneva (as the CERN site was still under construction).

About 300 physicists attended the meeting, including a sizable delegation from Russia. The three MURA representatives gave six talks on different aspects of FFAG accelerators [Kerst, 1956; Jones, 1956A; Symon, 1956B; Symon, 1956C]. Terwilliger, Laslett, and Sessler were coauthors of some of these papers. The contribution of Symon and Sessler was not accepted for oral presentation, but Kerst withdrew one of his contributions (which had been accepted), so as to provide a slot for the rf theory paper. Many years later, the CERN people were still apologizing to Sessler for their error of judgment as to the significance of his 1956 contribution.

The discussions following the FFAG presentations evidenced significant interest by the attendees. Kolomenski (from the Lebedev Institute) noted that he had independently conceived of FFAG, and suggested that the circumference factor could be reduced by operating on an "island" on the stability diagram, where the radial oscillations experience a phase advance of over half a wavelength in the radially focusing magnets. This would lead to a much smaller circumference factor. Jones replied that the MURA group had also noted this possibility, and, in the MURA nomenclature, it is referred to as Mark III. However, it is not an attractive option, because the tolerance on the machine parameters (e.g., the field gradient) is at least an order of magnitude greater than in the Mark I and Mark V designs (to avoid resonances) [Jones, 1954]. After Kerst's discussion of colliding beams, Rolf Wideröe rose to point out that he had earlier invented colliding beams and had proposed a collider geometry wherein protons would circulate in opposite directions in a ring using electrostatic fields to bend and focus the beams. Following his remarks, Ernest Lawrence delivered a devastating putdown, pointing out the impracticality of the electrostatic containment. The proceedings of the symposium record this rebuttal only very briefly and make it much milder than was in fact the case.

One evening the MURA group had a pleasant dinner with I. I. Rabi, who had been instrumental in the founding of CERN, and W. K. Jentschke, then on the faculty at Illinois but one of the founders and later director of the DESY laboratory in Germany, at a nice restaurant, the Pied de Cochon in the "Old Town" — the center of Geneva. A memorable event during this conference was a reception, held at Hotel Métropole (a very elegant Geneva hotel facing the lake) and hosted by the Russian delegation. Fred Mills, who later spent a year at Saclay Laboratory in France, noted that the senior French accelerator physicist H. Bruck had once recalled this symposium as an intellectual duel between the MURA and the Soviet physicists. Hildred Blewett wrote a nice summary of this conference for *Physics Today* [Blewett, 1956].

Tihiro Ohkawa from Japan had joined the MURA group in the autumn of 1955 on the basis of an invitation from Kerst, precipitated by learning that Ohkawa had independently invented the FFAG concept. One of his original ideas was the two-way Mark Ia colliding beam accelerator. Ohkawa published the concept in 1958 [Ohkawa, 1958]. Jones worked with him (primarily to help him with his English) to make a presentation of this idea at the 1957 Annual Physical Society Meeting in New York, where the idea seemed to be very well received by the media, and was cited in articles in the *New York Times* and *Time* magazine.

In August 1957, a small workshop was held at Berkeley to discuss an idea suggested by Budker that very intense, moderate energy circulating electron beams might pinch down to provide a very strong local magnetic field which might serve as the guide field for multi-GeV protons. Jones, Laslett, and Ohkawa (from MURA) participated in this workshop with others, including Dave Judd, Ernest Courant, and Lloyd Smith. Jones summarized some of these discussions in a MURA report [Jones, 1957C].

In 1959, a second major meeting was held at CERN — the International Conference on High Energy Accelerators and Instrumentation. This time, a large delegation of MURA physicists attended: Frank Cole, Robert Haxby, Lawrence Jones, Jackson Laslett, Fred Mills, Andrew Sessler, Keith Symon, and Kent Terwilliger (Fig. 3.20). Ohkawa was also there, although he had moved back to Japan, and his

Fig. 3.20. MURA representatives at the 1959 CERN Accelerator Conference. At the center of the picture, L-R, are Fred Mills, Nils Vogt-Nilsen and Frank Cole.

address was listed there as the University of Tokyo. A dozen papers were presented by the MURA representatives, eight of them in the first major session, on "Advances in High-Energy Particle Accelerators." Results from the rf studies with the Radial Sector (Michigan) Model, the studies with the Spiral Sector (Illinois) Model, the design of the two-way, 50 MeV Wisconsin Model, and other design studies were presented, including Terwilliger's concept of increasing luminosity by the superposition of equilibrium orbits of different energies in the collider straight sections [Jones, 1959B; Cole, 1959A; Laslett, 1959; Terwilliger, 1959; Jones, 1959c, 1959D; Symon, 1959A; Haxby, 1959; Cole, 1959B; Nielsen, 1959A; Meier, 1959]. A pleasant social function during this meeting was a dinner cruise on Lake Geneva (Fig. 3.21).

In 1961, the International Conference on High-Energy Accelerators was held at Brookhaven. Participants from MURA (or closely associated with MURA) included Frank Cole, Bob Haxby, L. Jackson Laslett, Fred Mills, Carl Nielsen, Nils Vogt-Nilsen, George Parzen, Don Swenson, Bernie Waldman, and Bill Wallenmeyer. The MURA staff presented seven papers at this meeting, on the design of a high-energy FFAG machine, theoretical studies of orbits, magnet design studies, etc., as cited in the bibliography [Waldman, 1961; Barton, 1961; Swenson, 1961; Cole, 1961; Mills, 1961A; Mills, 1961B; Pentz, 1961; Haxby, 1961; Parzen, 1961].

In 1963, the Russians organized an international conference on high energy accelerators. At Madison, an informal class in Russian was conducted, with a University of Wisconsin instructor, so the MURA delegates could at least read the Cyrillic alphabet, and say "Hello," "Thank you," etc. This was in the midst of the

Fig. 3.21. A boat cruise on Lake Geneva during the 1959 CERN Accelerator Conference. Standing, L-R: M. Seidl, Milton White, Bas de Raad, Fred Mills, and a waiter. Seated, L-R: Kent Terwilliger, Nils Vogt-Nilsen, an unidentified woman, and Frank Cole.

Cold War, and travel to Russia was very unusual. The Americans who were invited to attend this meeting were invited to Washington for a briefing on Russia and were given English maps of Moscow (which proved very useful). The conference was at Dubna, the site of the Russian 10 GeV proton synchrotron ("synchro-Phasotron"), a 680 MeV synchrocyclotron, and other accelerator work. The Americans first flew to Moscow and spent a few days there, where they saw the Kremlin and other tourist attractions, and visited the Institute for Experimental and Theoretical Physics (ITEP), where Alikhanov had built the electron accelerator model of the Dubna accelerator. (In America, a model of the Berkeley Bevatron was made and shipped to Caltech, where it was developed as a GeV electron synchrotron — a very analogous accelerator prototype development.) The Dubna site, on the shore of the River Volga, is about 117 km north of Moscow, and provided a very pleasant rural venue for the meeting. A cruise on the Volga was the nonconference highlight of the Dubna visit.

The MURA group presented seven papers at this meeting, including two discussing the MURA 50 MeV electron machine and one on proton linear accelerator design. Earlier in the summer, Jones had been a part of a Brookhaven summer study on the possibility of constructing colliding beam storage rings at Brookhaven. However, as this study was in progress, the Brookhaven administration leaned away from the colliding beams idea in favor of using their (always limited) resources to build a new linear accelerator injector, to enable a higher-intensity beam from the

30 GeV AGS. Perhaps this is why Jones, having no official connection with Brookhaven, was selected to present the results of this colliding beams study at the Dubna meeting. One original item of his presentation was the first discussion on the concept of detectors within the beam pipe of an accelerator for studying elastic scattering, etc. These later became known as "Roman pots," as an Italian group implemented this concept at the CERN Intersecting Storage Ring. The MURA representatives at this meeting were Larry Jones, Jackson Laslett, Fred Mills, George Parzen, Stan Snowdon, Keith Symon, and Don Young. The papers presented are cited in the bibliography [Jones, 1963A; Laslett, 1963A; Mills, 1963A; Mills, 1963B; Parzen, 1963; Snowdon, 1963; Young, 1963].

G. I. (Gersh Itskovich) Budker, who was at the conference, invited a group of Westerners, including members of the MURA delegation, to visit his new Institute of Nuclear Physics and its laboratories at the Akademgorodok ("Academic Town") near Novosibirsk in Siberia, following the conference — and this was indeed a most interesting journey! (Fig. 3.22). This visit was arranged at the conference. There was no Intourist office in Novosibirsk, as this area was, at that time, officially closed to Americans. The group included six Europeans and the seven Americans noted below. The tour was organized into two groups with visits of three days, with overnight flights from and to Moscow, although the second group had to spend the night *en route* to Novosibirsk in Omsk, due to bad weather in Novosibirsk.

Hildred Blewett was deposited in the Hotel of the Little Golden Valley. Exhausted, she decided to take a warm bath before sleeping. Much to her chagrin,

Fig. 3.22. Gersh Budker and Fred Mills seated at the Round Table in Novosibirsk in 1976.

no water was forthcoming from the faucets. Unable to communicate with the management, she went to bed and slept soundly. Somewhat later, the water was turned back on, filled the tub, and overflowed into the room and hall while she slept on. After a lot of handwringing and yelling, the hotel staff managed to wake her up and turn off the water and then move her to another room.

The two groups overlapped for half a day. These were the first Western physicists to visit the Akademgorod (although they were preceded by 23 US mathematicians, who had just left). The first group, who were featured on local television, included Fred Mills, Gerry O'Neill of Princeton, Arnold Schoch of CERN, plus John and Hildred Blewett of Brookhaven. The second group, who came following a post-conference visit to Leningrad, included Keith Symon plus Larry Jones and his wife, Ruth. Budker hosted the group very well, showing them his various laboratory construction and operational activities, and held interesting discussions. They saw the synchrotron radiation from the electron–electron collider, VEPP-1. They sat at Budker's "Round Table" and drank coffee in the Institute. This round table — a very large table — had seating for the principal physicists in the Institute and was used every day for coffee and discussions of the Institute's programs. This "democratic idea" was foreign to other Soviet establishments; it was most novel, and clearly very effective, as Budker's institute was one of the best in the Soviet Union. The Western visitors were also invited to Budker's home for dinner, where they met Budker's daughter and son-in-law, Max Zolotorev (who later emigrated to the US). Max was the architect of the experiment which discovered parity violation in atomic transitions, which assured the Nobel Prize for Weinberg and Salam (but not for him). Among the other Russian physicists there were R. Sagdeev, V. A. Sidorov, S. N. Rodionov, G. I. Dimov, V. L. Auslender, I. N. Meshkov, and B. V. Chirikov.

Jones' wife Ruth (who had joined him in Dubna and also came on this journey) and Hildred Blewett may have been the first Western women (other than Pat Nixon) to visit this part of Siberia since WWII. The visitors also were treated to an outing including a boat ride and picnic on the Ob Sea (formed by a dam on the Ob River). In walks with staff members around the Akademgorod, they were always followed by a limping man with a cane, who nevertheless always remained within earshot of the group. The second group was led from Moscow to Novosibirsk by an Intourist guide, Ludmilla Baranova, who obviously knew Budker quite well. She subsequently married Budker, and later accompanied him and Skrinsky on a visit to the US, including Madison. At their departure from Novosibirsk, Ruth was presented with a balalaika by Budker, which had been his son-in-law's, and which she subsequently learned to play.

In 1964, the biennial (Rochester) International Conference on High Energy Physics was also held at Dubna. It was at this meeting that, in discussions with Cocconi, Reines, and others, Jones became interested in studying high-energy elementary particle interactions with cosmic rays, while awaiting the construction of an accelerator (or storage rings) with energies above 100 GeV. These discussions also resulted in the conference at the Case Institute in Cleveland that autumn, at which Jones, Reines, and Cocconi, joined by L. Alvarez, R. Thompson, J. Keuffel, Y. Pal, S. Neddermeyer, A. Pevsner, M. Schwartz, F. Mills, and others, further discussed the exploitation of cosmic rays for basic particle physics research (discussed in Ch. 5, Sec. 5.5).

During July 20–24, 1964, a conference on linear accelerators was held at the MURA Laboratory at Stoughton. It was international in character, with 96 attendees representing 24 institutions from 7 countries in North America, Europe, and Asia. Although the focus of the conference was primarily on proton linacs, it was also an opportune time for electron linacs. The new "monster" linac at the Stanford Linear Accelerator Center (SLAC) was approaching completion, and there was serious discussion of potential instabilities, which indeed appeared later in SLAC operation. The highlight of the conference, however, was the great improvement in rf cavity design and orbit design achieved by the MURA group using digital computation. By this time, the design methods used at MURA were the accepted methods to employ, and, as noted elsewhere (Sec. 5.9), were used to design proton linacs built at LANL, BNL, and FNAL. Altogether 54 papers were presented, 6 by MURA participants [Young, 1964; Kriegler, 1964; Swenson, 1964A; Curtis, 1964; Swenson, 1964B; O'Meara, 1964].

The conference was organized, and the proceedings published, by a committee consisting of C. D. Curtis, F. E. Mills (Chair), D. A. Swenson, and D. E. Young. The conference was the fourth in a series that is still held on alternate years, and is now truly international. D. E. Young, in 1970, chaired the conference held at FNAL.

After the 1963, Novosibirsk visit, Budker had expressed his strong desire for an annual meeting of physicists engaged in colliding beam work. This is reflected in a letter to 15 US physicists (including those who visited Novosibirsk in 1963) by Gerry O'Neill, noting Budker's desire for a meeting in either Novosibirsk or Dubna, following this 1964 Dubna Conference. That postconference meeting did not occur, however. But in March 1965, Budker invited a number of Americans (about ten) to come to Novosibirsk to discuss colliding beams at the "Colliding Beams Section" of the International Conference on Multiparticle Interactions. Included were Bruno Touschek, Andy Sessler, Ernest Courant, Gerry and Vera (Kistiakowsky) Fisher, Burt Richter, Bernie Gittelman, and Fred Mills. Andrei Lebedev and Sergei Kapitsa from Moscow showed up briefly, but they were there for Skrinsky's thesis defense. They met for about two weeks, discussing all of the problems — both physics and engineering — which remained to be overcome. At that time, each was assigned a Soviet to "look after him"; Sessler was assigned Skrinsky, who was then just completing his Ph.D. program. Skrinsky carefully looked out for Sessler's welfare, escorting him, helping him get lunch, etc. On Sunday they went cross-country skiing, at which Sessler was a novice. Most of the afternoon went well, but later Sessler hit a tree and broke the tip of his ski, which he retains as a souvenir of Siberia. Sessler also met Boris Chirikov at that time, who told him about Chirikov's criterion before he had actually published it. Their friendship continued and resulted in a joint paper. Chirikov thought that it would be good to get a German involved and so, perhaps, to have one of the first papers published after WWII with a Russian, a German, and an American as coauthors. They got Eberhard Keil to join them in this. (Unfortunately they were scooped by a 1948 paper by Alpher, Bethe, and Gamow.)

A summer study at SLAC was organized for about a month in 1965, devoted to colliding beam problems, and it included Sessler. They attacked and solved many of the problems identified at the earlier meeting in Siberia. Sessler and Claudio Pellegrini met at that study, and coauthored a paper, starting a friendship and collaboration that resulted in many more papers and a lasting cordial relationship.

In the late summer of 1965, the International Accelerator Conference was held in Frascati, Italy. Jones was at CERN during that year, and wished to present his ideas on the use of cosmic rays to explore 100 GeV–TeV proton–proton interactions in the absence of accelerators of those energies. The biennial International Cosmic Ray Conference was in London in 1965, but, as he had no recognition in the cosmic ray community at that time, Jones was unable to obtain an invitation to that conference. However, he was well known in the accelerator community, and hence gave a lecture on his cosmic ray ideas at the Frascati accelerator conference. Eduardo Amaldi gave a summary lecture at this meeting, and generally rejected Jones' proposals, arguing that appropriate accelerator facilities would indeed be built in the near future. Sessler was also present, and participated in a memorable round-table. Each of the participants, perhaps eight, was seated with a full bottle of Frascati white wine in front of him. Bruno Touschek (who probably arranged for the wine) was the chair. It was very hot that afternoon in the conference facility, and the discussion very soon showed the effect of the wine. Speaking of drinking, about halfway through the conference, the Americans discovered that the price of wine was less than that of milk, and indeed some indulged to excess.

At another afternoon parallel session chaired by Matt Sands, Ken Green mumbled under his breath so audibly and consistently that Sands asked "the parallel session to the parallel session to please leave." The meeting was also attended by Mills, Sessler, Snowdon, Symon, and Young. The papers presented reflected the changing interests of the MURA group, and the changing mission of the MURA Laboratory.

One afternoon Jones and Mills took a bus to Rome to adjust their airline reservations with Pan American Airlines. As the bus penetrated further and further into Rome, the traffic became heavier and heavier. Indeed, cars were even being driven on the sidewalks. Soon the bus stopped, mired in a crowd of about a million people. They had accidentally happened upon a great event where the Pope went in procession from San Giovanni del Laterno to the Vatican. Stopping to watch, they soon became aware of a short, fat Italian boy and his mother. The boy was distressed because he could not see "*Il Papa.*" In obeisance to the mother's pleading eyes, they hoisted the boy to their shoulders for him to look. When the procession passed, they put the boy down and he and his mother disappeared into the crowd, no words having been spoken.

In 1966, Sessler and Rowe attended the International Symposium on Electron and Positron Storage Rings at Saclay — the first conference devoted solely to this subject, with very many relevant papers and excellent communication among the international community. Sessler was at CERN for the year, and deeply involved with the ISR and all other aspects of colliding beam devices; Rowe, at MURA, was midway into the construction of the 240 MeV storage ring there. Although the MURA (and ex-MURA) physicists continued to participate in significant national and international conferences on accelerator-related matters, these meetings in the 1950s and 1960s are most relevant to this historical discussion.

Chapter 4

THE MADISON YEARS, 1956–1963

4.1. FORMATION OF THE MURA ORGANIZATION

During 1953–1954, the Midwest accelerator design and discussion activities were carried out by an organization called the "Midwest Accelerator Conference" (MAC). This was a rather informal organization of the physicists from the several large Midwestern state universities who were interested in supporting high-energy particle physics in this area. This group of senior physicists included Gerald Kruger (Illinois), Dick Crane (Michigan), John Williams (Minnesota), Ragnar Rollefson (Wisconsin), Bob Haxby (Purdue), Alan Mitchell (Indiana), Josef Jauch (Iowa), and others. The modest government grants to support this research were mostly from the existing AEC and/or NSF programs at the universities, in particular Illinois (Kerst's school, and the site of major computer facilities). Following the summer of 1954, funds from AEC were not forthcoming, and the group turned to NSF and the Office of Naval Research (which, at that time, still supported significant basic physics research), although the major support still came from the participating universities.

During the fall of 1954, these senior physicists became the organizing committee for a new, more formally chartered organization: MURA, a State of Illinois corporation. Henceforth, federal funds were solicited by and came through MURA. The universities of MURA were primarily the "Big Ten," or Committee on Institutional Cooperation, universities (of intercollegiate athletics note). The 15 member schools were the University of Kansas, Iowa State University, University of Iowa, University of Minnesota, Washington University (St. Louis), University of Illinois, University of Chicago, Northwestern University, University of Wisconsin, Michigan State University, University of Michigan, Notre Dame University, Purdue University, University of Indiana, and Ohio State University. Each university contributed US$ 15,000 to the MURA treasury to enable the organization to get started. Later, the MURA technical reports would all have a cover page that included a simple map of the Midwest region (Fig. 4.1). With a small pennant on a pole showing the initials and locations of each university, this design came to be

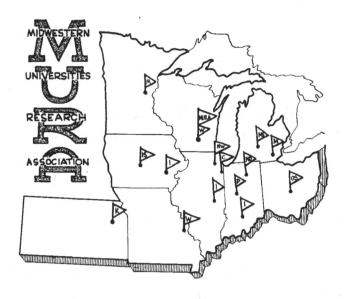

CONSTANT FREQUENCY CYCLOTRONS WITH SPIRALLY RIDGED POLES

D. W. Kerst

REPORT   NUMBER 64

Fig. 4.1.   The standard cover of MURA technical reports, "The Golf Course": a map of the Midwest with MURA universities noted with flags.

called the "Golf Course" within MURA. The MURA Corporation, with oversight from a governing board representing the senior physicists at the member universities, was a good administrative structure, and successfully managed the activities of the working group of physicists, who (except for Kerst) were mostly younger.

4.2. THE MOVE TO MADISON

As noted in Sec. 3.3, a strong and active working group had formed at Urbana in 1955; during that autumn (following the Michigan summer study) the group included Cole, Sessler, Symon, Laslett, Jim Snyder, Nils Vogt-Nilson (from Norway), and Edward Akeley (who commuted from Purdue), all under Kerst's leadership. Jones and Terwilliger were busy constructing the Radial Sector Model (Sec. 3.6), and hence remained at Ann Arbor, though their frequent visits to Urbana included the periodic meetings of the large group.

In 1956, MURA decided that it needed an identifiable working location, separate from the limited space borrowed from an existing physics department (the office space on the Illinois campus at Urbana had become very crowded indeed), and also sought a site for the multi-GeV accelerator which it hoped to build, supported and funded by the government. The different MURA universities proposed possible sites, and the MURA board decided on a site near the University of Wisconsin for its laboratory. An area of farmland a few miles south of Madison, near the town of Stoughton, was identified as a potential site for a major accelerator, and a future central laboratory. However, until more significant funding was available and (perhaps) a major accelerator proposal was accepted, it was decided to find a location in Madison, not too far from the University. So, in mid-1956, MURA rented a property on University Avenue, about a mile west of the Madison campus. The building had been a Nash automobile dealership and garage; the Nash automobile company had recently folded, so the building had been vacated. This "Nash Garage," as the MURA group always called it, was quite appropriate for the needs of the MURA working group at that time. The front (street side; former show room) provided space for group meetings and administrative offices plus other offices. The second floor had abundant additional office space for the active MURA physicists. Behind these areas, the repair garage provided adequate floor space for laboratory activities, as well as computers, shop space, and (most importantly) space for the Radial and Spiral Models.

The move to the Nash Garage occurred during the summer and early fall of 1956. Kerst persuaded Jones and Terwilliger to move the Radial Sector Model from Ann Arbor, and to move (with their families) to Madison; this they did in September 1956, when they joined the larger group which had moved up from Urbana. The Radial Sector (Michigan) Model was reassembled quite efficiently, and the program of work on rf acceleration, phase displacement, etc. was pursued there (see Fig. 4.2). Charles Pruett, who had joined the Michigan group to work on the model, also came to Madison (with the model), where he remained throughout his career. The Spiral Sector Model, which had been designed, and its magnets constructed, at Illinois, was assembled and operated in the Nash Garage.

Following the summer of 1957, Jones and Terwilliger moved back to Michigan, to resume their academic careers, although remaining very active in MURA work; they commuted often to Madison for meetings and studies, as well as involvement in summer studies, workshops, and conferences. They argued that they had become involved in MURA to help to create a new, higher-energy machine on which they could pursue their first interest: experimental research in elementary particle physics. They also valued and sought careers as university professors. Back at Michigan, Terwilliger became involved with Donald Meyer in the development of the spark chamber, while Jones teamed up with Martin Perl to develop the scintillation chamber. Although remaining close friends, Jones and Terwilliger felt that they each needed to establish an independent identity; they had collaborated in

Fig. 4.2. Visit to the MURA laboratory in Madison by Niels Bohr in 1958. Viewing the Radial Sector Model are L-R: Robert Haxby, Ragnar Rollefson, Harrison Randall, Subramanyan Chandrasekhar, Niels Bohr, Charles Pruett, and Lawrence Jones.

graduate school on their Ph.D. theses, and all of their research activities had been joint efforts, resulting in joint publications.

Don Glaser hired Meyer and Perl, at Michigan, to help with the first physics experiments with the (then new) bubble chamber. This they did, but they had a bit of a falling-out with Glaser and so decided to find other collaborators. Among the doctoral students of Jones and Perl was Samuel Ting. Later, Glaser, Perl, and Ting would received Nobel Prizes in physics for their separate achievements.

During the subsequent academic year (1957–1958), the 50 MeV electron Two-Way (Wisconsin) Model was designed and construction began. To house the accelerator in the Nash Garage would have used up much of the available laboratory space and required extensive shielding for the other areas in and near the laboratory (a residence was located on the other side of the south wall of the building). However, it required more space than was available in the garage. Instead, the accelerator was constructed and assembled during 1958 and 1959 in an underground building built for the purpose in a steep little hill near the northeast corner of the Stoughton site. A prefabricated steel frame building for the associated control room, power supplies, and laboratory space was built on top of the hill. To house the burgeoning population of MURA employees, a building across University Avenue, the site of a former lumber store, was rented and refurbished with a library, offices, and a large seminar room.

A most significant period of the MURA residence in the Nash Garage was the 1959 Summer Workshop (see Sec. 4.6).

A few years later, a new laboratory building to house the MURA work was built at the Stoughton site. It was completed in the spring of 1963, and MURA moved out of the Nash Garage and lumber store and into the new building. The building was financed by the MURA Corporation. After extensive negotiation, AEC grudgingly agreed to pay a rental fee to the Corporation for its use for the research funded by AEC, just as they had agreed to pay a rental fee for use of the MURA 704 computer. This negotiation was a harbinger of things to come, for AEC believed that the income from the rental to non-MURA users of the 704 computer, purchased by the Corporation, rightfully belonged to the AEC, even though the Commission's own rules forbade it from owning computers. This disagreement would prolong the life of the Corporation by more than five years.

4.3. SPACE CHARGE

Prior to about 1956, there was essentially no work done on "space charge" effects — the influence of one particle upon another — in high-energy accelerators and, particularly, in AG accelerators. Accelerator builders were happy simply to get a few particles to high energy and were not concerned about the limitations that space charge effects entail. Of course, there had been work on gas focusing and, most importantly, the calutrons (used during WWII to separate the uranium isotope ^{235}U for the atomic bomb) only produced ^{235}U in adequate amounts because of clever compensation of space charge effects. However, there had been no work on space charge effects in high-energy accelerators.

Perhaps the first realization of the importance of space charge effects was Don Kerst's realization that injection into a betatron, to the degree observed, was due to self-forces [Kerst, 1948]. In 1956, at the First International Accelerator Conference, there were two very interesting contributions — one by Veksler [Veksler, 1956] and one by Budker [Budker, 1956], both of whose machines depended upon self-forces for their very operation. Neither of these ideas involved space charge effects in conventional accelerators, but they served to stimulate accelerator physicists.

Stimulated by the idea of colliding beams and, hence, the need for high current, and very much influenced by Kerst's experience with the betatron, the MURA group began to think about, and analyze, the effects of space charge in conventional accelerators. At first, a one-dimensional problem was addressed, namely the effect of self-forces on longitudinal distributions within rf buckets [Nielsen, 1959B]. This work soon led to a realization of a possible coherent instability, which is described in Sec. 3.10.

About the same time, thought was given to the effect on transverse motion. The first effort from the MURA group is described in two MURA reports

[MURA462; MURA466]. Lee Teng also devoted himself to the problem [Teng, 1960]. In addition, pieces were being developed by Fred Mills and Jean van Bladel, and there were old works in texts dating back to the 1920s. Soon, however, Jackson Laslett got involved and his definitive work, built upon the work of others, but carefully — as was his manner — considering all the terms involved, became the definitive paper on local transverse space charge effects [Laslett, 1963A,B]. To this day, that paper is quoted and the formulas within it used.

Three years later, at the Second International Accelerator Conference in 1959, there were many papers on the effect of particles upon each other. There were theoretical advances in the study of coherent instabilities, introduction of the "water bag model," and proposals for a plasma betatron and a plasma linear accelerator. The water bag model, first introduced by the MURA group in a one-dimensional application, was used to study transverse space charge limits in a linac by Kapchinskij and Vladimirskij [Kapchinskij, 1959]. It is a model in which the phase space density is taken to be uniform and constant within a certain boundary, so that, because of Liouville's theorem, only the motion of the boundary needs to be followed. (In one dimension, this means describing how a simple curve's shape changes as the motion proceeds.)

In summary, then, subsequent to the MURA work, much research was done on collective instabilities. In fact, through the years, many scientists have devoted their professional lives to the study and cure of collective instabilities. The incoherent space charge limit (which is what Laslett evaluated) has been employed over and over again. Most particularly, in a colliding beam machine, the incoherent interaction of a particle with the field of the opposite beam is the limit in performance. A major advance was the realization that there can be nonlocal effects. Charges may leave behind fields that stay at a particular position in the accelerator and they may also have fields that follow their passage, such as in a wake. Once again, the study of such phenomena is a major activity of many accelerator physicists.

4.4. INJECTION AND EXTRACTION

The nature of orbits in circular (and cyclic) particle accelerators is such that any particle will pass repeatedly around the accelerator oscillating horizontally and vertically about some fixed path in space called its equilibrium orbit (e.o.). The particle will do this as long as the magnetic guide field is constant in time and the particle does not collide with an atom, does not radiate, and is not subjected to a transient electromagnetic field that might deflect or accelerate it.

The earliest circular accelerators, so-called "weak focusing" accelerators, had large "dispersion," which is to say that the position of their e.o. changed rapidly as the particle energy changed. At the same time, for a fixed energy, a change in the magnetic guide field caused a large change in the position of the e.o. If the rate of change of the guide field was small, the amplitude of oscillation around the moving e.o. remained constant, so the particle followed its e.o. This allowed the injection and accumulation of many turns of beam from the injector, and an increase in the number of particles accelerated. On the other hand, a rapid change of energy, caused, for example, by passing through a piece of matter, caused an instantaneous change in the position of a particle's e.o. These properties of high dispersion were exploited in early proton synchrotrons to facilitate injection into and extraction from the synchrotrons (Sec. 3.9).

When AG or "strong focusing" (low dispersion) accelerators were built, those methods were no longer useful. A transient field was used to inject a single turn of beam into the first AG proton accelerators — the Brookhaven Alternating Gradient Synchrotron (AGS) and the CERN Proton Synchrotron (PS). The beams were not extracted from these accelerators; rather, they employed internal targets to make secondary beams, which were used for experimentation.

At MURA, the problem of injection was addressed first by Kerst and Mills, and then by Christian and others [Christian, 1961; Kerst, 1957; Mills, 1959; Mills, 1961A; Mills, 1961B]. A method of transverse stacking was devised which consisted of slowly turning off a transient deflection system that moved the orbits away from the injection structures, thus mimicking, in a sense, the methods using large dispersion. This method, tested in the MURA models, was used in the AGS for many years.

In the case of extraction, a method which in a sense is the inverse of the injection process above, the "unstacking" was required. However, the need was to unstack over long times, many turns, with high efficiency. The solution was presented by Symon [Laslett, 1959B], and consisted of adding field components whose geometrical arrangement is in resonance with the particles' oscillatory motion, so that it is possible to cause the oscillations to grow. If the field components are nonlinear, then it is possible to make a situation in which small oscillations are stable, and large oscillations grow rapidly, even exponentially. Symon's idea was to create a stable region in phase space to contain the beam, which region then can be made to decrease in area. By increasing the strength of these field components slowly, all oscillation amplitudes eventually become unstable and can be made to exit the accelerator in a continuous and controlled way. Laslett collaborated on the development of this scheme, particularly in numerical calculations and in working out the phase curves. This method was extremely successful and was employed on the Berkeley Bevatron, the AGS, the PS, and the CERN Super Proton Synchrotron (SPS), among others, using sextupole magnets as the added field components. At Fermilab, a similar scheme was employed using quadrupole and octupole magnets. The first resonant extraction scheme was used on the betatron at the University of Illinois [Skaggs, 1946]. An adaptation of this extraction scheme using betatron resonances was proposed by Tuck and Teng [Tuck, 1950]. However, the work described here was probably the first to make explicit use of slow variation of invariant phase curves. It is noted in the paper cited above [Laslett, 1959B] that, for this kind of scheme, time reversal turns an extraction scheme into an injection scheme. Thus, the paper discusses both extraction and injection [Laslett, 1959B]. Although this observation is correct, it does not follow that one can fashion a successful injection scheme from it. In such a system, the horizontal phase space is leaked out over many turns, so that the extracted beam horizontal emittance is much smaller than that of the circulating beam. The vertical emittance is unchanged by the process. On the other hand, most useful injectors provide beams with comparable horizontal and vertical emittances. Then the only way the system can work in reverse is to start with an incredibly small emittance, something not so easy to find. The paper in the 1959 CERN Symposium by Jones and Terwilliger [Jones, 1959D] discusses a number of extraction schemes.

It is interesting to note that resonance extraction had been used earlier in the 22 MeV betatron at the University of Illinois, and in cyclotrons at the University of Chicago, but these systems failed to allow the slow controlled extraction needed by the physics program at synchrotrons.

4.5. THE 50 MeV TWO-WAY MODEL

Motivation

Experience with the small Radial Sector and Spiral Sector Models was positive and encouraging, but many problems had not been addressed. On the construction side, the magnets had been of the scaling pole type, which does not lend itself to producing the high fields necessary to reach high energies; thus nonscaling pole magnets were needed. Second, the low dispersion and low energies of the models made injection easy using expander coils in the radial sector, or extremely thin septa in the spiral to accomplish injection with small accelerating voltages. Usable rf cavities were not developed which would reliably accelerate the beam over large radial spans. Vacuum adequate for beam storage had not been demonstrated.

On the side of beam dynamics, high-intensity beams, particularly in the stacking mode, were not studied. Although proof-of-principle studies of beam stacking and transition crossing had been performed, these studies were not of sufficient depth to move ahead. At least one instability, the negative mass instability, would stand in the way of success. Other instabilities were suspected, as well as limitations due to self-focusing of the beam owing to its space charge and ions trapped in the beam.

Tihiro Ohkawa came up with the beautiful idea of designing an accelerator such that particles of the same sign would travel both clockwise and counter clockwise and thus be able to undergo collision. It was realized that a new model could be built with this feature and at the same time address many of the issues described above both on the construction and beam dynamics side. Thus, MURA initiated the 50 MeV Two-Way (Wisconsin) Model [MURA, 1964].

Lattice Design

Interest had been shown in the two-way design proposed by Ohkawa, particularly as a possible way to study colliding beams. It was felt that many of the goals of the next model might be achieved in that configuration. At the same time, if problems were encountered, a suitable one-way operating point had to be available. The choice made, $k = 9.25$, $N = 16$ with sufficient space between magnets for rf cavities, insulating gaps, and structural support for the vacuum chamber, led to horizontal and vertical tunes of 6.36 and 5.34. These were uncomfortably high, with phase advances per sector of 0.80π and 0.67π. The beat factors (ratio of max to min beta functions) were large. The primary reason for this was the large value of k, which was necessary for keeping the radial aperture acceptable. A magnet for the Two-Way Model is shown in Fig. 4.3 (the completed accelerator is shown in Fig. 1.3).

Magnets

Magnets for the previous models were constructed of poles (surfaces of steel) lying in cones, and terminated azimuthally on rays that were radial in the Radial Sector Model and spiral in the Spiral Sector Model. Low-current windings excited the poles, and the field was adjusted to have the proper radial dependence by returning

Fig. 4.3.   A magnet of the 50 MeV Model.

some of the current across the pole to adjust the magnetic potential. Fields were in the range of 100 G only. To reach an energy where the beam would have a sufficient lifetime required stronger fields, higher currents, and different methods. The 32 identical magnets used simple back winding on the surface of the iron pole at inner radii; then, as the current per unit radius increased, the windings were located in slots in the poles. The pole region between slots was shaped to lie on a magnetic equipotential surface appropriate to its location. When the current per slot became too high, back winding was abandoned in favor of a three-dimensional pole surface, the "nonscaling pole," which was to provide the proper field shape at the higher fields. A view of one of the 32 magnets is seen in Fig. 4.3. Three-dimensional magnetic field programs were only a dream to be realized 35 years later, so an approximate method was employed which used two-dimensional field programs. And, finally, tuning windings were installed to change tunes. A solid conductor was used, with water-cooled "aprons" embedded in the windings.

Magnet Measurements

The corrections to the guide field magnets occurred in two separate phases. The first phase was done before two-way operation of the accelerator and was concerned mainly with producing a uniform value of k. After the difficulties in the magnetic fields were realized during the operation of the accelerator in the two-way mode, major changes were made in the measuring instruments and the interpretation of the measurement data before attempting to operate in the one-way mode. In this second stage, new computer programs and improved measuring instruments and methods were developed for correcting the fields. The measuring instruments provided a way of accurate mapping of the fields, and the computer programs allowed rapid reduction of the data to useful form and the detection of excessive magnitudes of undesirable harmonics in the field. Further, the measurement of fields on a mesh allowed the use of established Runge–Kutta techniques to calculate orbits directly from the data. The calculation predicted the slight wandering of the tune and predicted the presence of several nonlinear imperfection resonances. In operation, these resonances evidenced themselves by measurement of lifetime versus momentum. The stacking region was chosen to avoid these resonances.

The goal of making corrections to the magnetic field for operation in the one-way mode was also to allow operation in the two-way mode. The main intent of the one-way corrections was to improve the high-intensity capability of the accelerator. With different operating points for the two modes of operation, this put severe requirements on the tolerances of the measurements. It was possible to develop correction shims and correction coils that required only changes in the currents in the coils to shift from the one-way operating point to the two-way operating point.

Vacuum

After exploration of high-temperature-baked nonmagnetic stainless steel chambers, and the lack of commercial ultrahigh-vacuum pumps or seals, it was decided to construct an aluminum chamber supported by I-beams welded to it, baked at moderate (150° C) temperatures, and pumped with prebaked vac-ion (getter ion) pumps. Four full radial span insulating gaps were provided — two for rf cavities and two for betatron acceleration. Insulation was provided by machined Teflon slabs that were shielded so that scattered electrons could not reach them. Vacuum seals throughout were of flouroelastomer (Viton), which could be baked at several hundred degrees C. Sliding seals, of which there were 18 with 40″ penetration, used two o-rings pumped between by forepumps equipped with baked charcoal traps. Molybdenum sulfide powder was used as lubricant.

After baking, the pressure reached 2×10^{-8} Torr. Following several failures (broken probes), water and argon increased the pressure to about 2×10^{-7} Torr.

Betatron Acceleration

Although some tests of transition crossing were being made, it was decided to avoid this problem in the model by accelerating to 2 MeV by betatron acceleration, capturing the beam, and accelerating to the stacking energy. In order to accommodate beams in two directions, on each cycle the cores received two pulses of opposite polarity during which the capacitances were recharged by tightly coupling a transformer winding into the circuit. Injection was timed by measuring the flux in the cores during the pulse and sending a pulse when the desired accelerating flux was present and the sign of the accelerating voltage was proper. One direction of beam was injected on the first pulse, after which the second pulse would accelerate and decelerate it while the second beam was being accelerated. Thus the two beams could be captured together with the rf system, accelerated, and stacked.

RF Cavities

Two cavities were required — one to accelerate the beam from 2 MeV to the stacking energies up to 45 MeV, and one to resupply the energy loss due to synchrotron radiation. The radiation energy loss per turn in the stacked beam is about 1 eV. If one were to try to make up for this small loss in the usual way by holding the stacked beam in an accelerating bucket, the bucket would have to be very large and hence would require a large gap voltage because of the energy spread of the stacked beam. Instead, in the Two-Way machine, a small empty bucket is decelerated down through the beam repeatedly to provide an average acceleration of about 1 eV per turn by phase displacement. The cavities consisted of an inductive rhumbatron coupled to the capacitance of the insulating gap of the chamber. This was loaded resistively to achieve the desired broadband operation. (The rhumbatron or rumbatron is an electromagnetic microwave cavity resonator, invented by W. W. Hansen at Stanford and thought to have gotten its name from the back-and-forth travel of electromagnetic waves inside the cavity.) This was loaded resistively to achieve the desired broadband operation. The broadband push–pull amplifier was coupled directly to the gap.

Injection

There were no existing electron guns that were readily adaptable to injection into this model. It was decided to utilize a modification of a commercial X-ray tube that operated in the same voltage range. This tube was purchased and tested, and was used on the early operation in 1959. The beam quality was measured, but was significantly worse than might be hoped, and was very astigmatic; in the crowded environs of the central region of the machine, there was not sufficient space to employ a beam transport system to match the beam into the accelerator. For the first studies, a simple solenoidal lens helped deliver some beam through the fringe field on the inside of a D magnet to an electrostatic inflector, which was sufficient for beam studies.

After the initial running period, a new "high field" gun was developed which showed little space charge effect up to currents of 250 mA (much more than was needed). The gun incorporated a focusing electrostatic inflector that brought the beam parallel to the injection equilibrium orbit. The whole was inserted into the vacuum chamber on a mount that allowed adjustment of position and angle. Obtaining space-charge-limited beam was routine with this gun. It was used to test multiturn injection techniques, but the thin septum of the inflector allowed sufficient beam to be injected by the betatron acceleration, or by space charge. A drawing of the gun-inflector system is seen in Fig. 4.4.

Operation

The first running, at the two-way operating point, showed several inadequacies. First, the nonscaling pole did not provide the needed zero chromaticity. In fact, the tune wandered greatly, crossing an integral resonance. Second, the original electron gun would not provide the needed current to obtain intensities desired. Third, the magnet power supply, a very early solid state system with variable inductors, did not have adequate stability. Fourth, the sequence control system was insufficient to create the stacking cycles desired. The tune profiles were measured, the rf systems tested, two-way acceleration tested, and the machine was shut down for improvements. Because of the lack of continued interest in colliding beams, and the growth

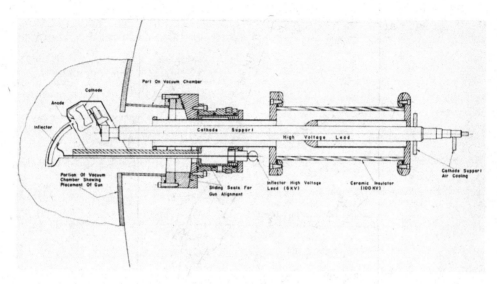

Fig. 4.4. The 50 MeV Model gun-inflector system. A converging beam of electrons emerges from a heated ribbon of tungsten located in the cathode electrode. The beam passes through a slot, where it is defocused, into the transverse field of the inflector, which deflects and focuses it to emerge into the main vacuum chamber in the correct direction. The whole assembly can be rotated to change the beam direction. The cathode is air cooled to avoid overheating. The system operated successfully at voltages up to 150 kV.

of interest in high intensities, it was decided to move the operating point to a one-way point that was considerably more conservative. (See Fig. 4.5.)

These improvements were completed 18 months later, and a second phase began. Over the following 24 months, the accelerator achieved a stacked beam of 10 A, in which the longitudinal phase space density was limited by the negative mass instability, the transverse phase space density was limited by space charge detuning, and the stacking efficiency exceeded 95%. The beam lifetime was determined by the growth of radial betatron oscillations due to synchrotron radiation. To achieve this result it had been necessary to remove ions from the beam by installing clearing electrodes, stabilize a multimode transverse instability with feedback, and stabilize the detuning of the injection field by the beam current. Essentially all original goals had been accomplished.

4.6. MURA PROPOSALS

From 1955 to 1963, MURA submitted a number of proposals to AEC for study and construction of a high-energy FFAG accelerator laboratory. Some of the proposals were a result of recommendations made by reviewers of previous proposals. Some

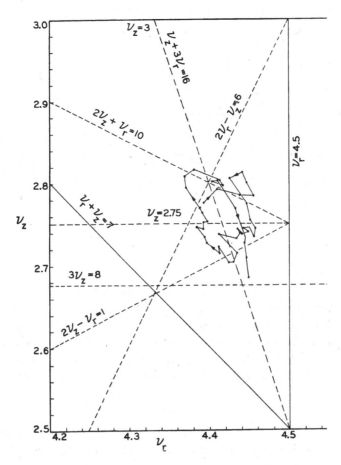

Fig. 4.5. Tune diagram for the 50 MeV Model at the one-way operating point. Here v_r is the horizontal tune, and v_z is the vertical tune of the betatron oscillations. One can observe the wandering of the tune during acceleration. Two visible resonances, $3v_r + v_z = 16$, and $v_r + 2v_z = 10$, caused beam loss when the beam was "parked" on them. These resonances were avoided in the stacking experiments.

arose from recommendations of various national advisory panels (see Sec. 5.8). Many of the MURA proposals were nicknamed after the color of their covers, as noted below.

April 15, 1955: Proposal to AEC for Cooperative Research in High Energy Physics, for the establishment of a laboratory in the Middle West which would serve the best educational and scientific needs of that area. The proposal included the construction of a 20 GeV Spiral Sector Synchrotron, for a proposed cost of US$ 9,256,000, to be completed in five years. The proposal also discussed a 10 GeV Spiral Sector FFAG accelerator and a 10 GeV Radial Sector FFAG accelerator. It was submitted by

P. G. Kruger, President; L. R. Lunden, Secretary; J. H. Williams, Vice President; A. W. Peterson, Treasurer; and D. W. Kerst, Director of the Technical Group.

April 5, 1956: Proposal to AEC for two 15 GeV high-intensity colliding beams for a total cost of US$ 100,000,000 (Brown Book). The proposal was produced by D. W. Kerst with help from P. G. Kruger. It was to make use of two tangent spiral sector rings of 15 GeV each, with one colliding beam area at the tangent point. It was a rather incomplete proposal. There was almost nothing in it about experimental devices or laboratory facilities, either for feasibility or for cost. The cost included US$ 75,000,000 for construction and US$ 25,000,000 for operation for ten years. The proposal included a long list of unanswered technical questions, with the statement that these all needed to be answered before the accelerator could be constructed.

On November 26, 1956, John Williams of Minnesota, the President of MURA, sent a telegram to AEC asking about the status of the proposal. The response was a long letter from Commissioner Vance, dated November 30. (Is it possible to imagine that nowadays any communication to the US government could receive a reasoned, thought-out answer in four days?) The Vance letter said:

(1) The proposal had neither been accepted nor rejected, but a technical review had been completed. The Commission had determined that there should be further studies to demonstrate the feasibility of FFAG and the usefulness of the accelerator for research. The Commission was prepared to support these studies. The MURA staff was complimented and hope was expressed that the studies would prove the merit of the facility.

(2) The site of the accelerator would not necessarily be at Argonne, because doubt had been cast in the proposal as to the geological suitability of Argonne. But the Commission took a firm position that there was no justification for a second large-scale laboratory unless geology demanded it. In the summing-up, it was stated that if the accelerator was built, construction and operation would be carried out as a fully integrated part of Argonne.

(3) AEC was not interested in a new management concept for Argonne, but desired to strengthen the ties between Argonne and the Midwestern universities. AEC was also not considering a change in the Argonne mission, which they regarded as 53% basic research and 47% applied research and development.

Wm. M. Brobeck & Associates reviewed the proposal in August 1958; they estimated a total construction cost of US$ 213,972,000. The final MURA estimate was US$ 110,500,000.

March 14, 1958: Proposal for two-way 15 GeV colliding beams and high intensity. Construction would take five years, and cost US$ 85,428,950. The design was based on the Ohkawa two-way geometry. The MURA group were bothered by the fact that the only design presented publicly for colliding beams was the incomplete one of the 1956 proposal. They did not want the state of their technical knowledge to be judged by it, so it was decided late in 1957 to work on this colliding beams proposal for a multi-GeV device. Large two-way rings suffer from a lack of vertical focusing, so the performance of the ring was not spectacular, but adequate. The group was able to state in the proposal that all the technical problems listed in the 1956 proposal had been solved and that these uncertainties were therefore behind them. They were very proud of that statement.

But there were new concerns. Firstly, due to recent theoretical work and, most importantly, experimental work at Stanford, the effect of one beam on the other beam had been observed and quantified. The electric and magnetic space charge forces in the two antiparallel colliding beams added, rather than subtracting as they would in parallel beams, and, even though an estimate was made, it was not clear how much disruption of the circulating beams there would be. Secondly, it was appreciated that there was very little room in the straight sections between magnets to put detectors for particle physics experiments. Thirdly, a detector with time resolution was needed to separate the desired beam–beam events from beam–background gas events, which were at least as likely as the desired events with the vacuum that could be achieved using the techniques of the day. In 1958, there was no such detector. MURA pointed out this lack in the document.

The proposal was submitted to AEC in the spring of 1958. They did not receive it very graciously; in fact, it was most unwelcome. When it was brought in, Paul MacDaniel, the Head of Physics Research, growled, "What do you want us to do with this?" The MURA reply was simply: "Send it out to the physics community for review." Some persuasion was required to get them to do what would nowadays be a virtually automatic response to any new proposal.

The reviews trickled in over the next several months. The general tenor of the reviews was that this was a good accelerator proposal, a large step forward from the 1956 one. But the reviewers were all troubled by the thought of spending money on an accelerator without a detector available. A large number of them commented that the available single high-intensity beams, included as a throw-in, would be valuable immediately and should be pursued. There had also been a report of a physics group to the President's Science Advisory Committee that held that all physics was asymptotic above 15 GeV, which nonsense helped to muddy the waters.

Later, a paper was published on the design to bring it before a wider audience [MURA, 1962].

March 30, 1962: Proposal for a single 10 GeV spiral sector high-intensity accelerator, to be built in seven years at a cost of US$ 79,014,00, and to produce 2×10^{14} protons per second, three orders of magnitude above what synchrotrons were producing at that time (Blue Book). It was a spiral sector ring with radial straight sections for rf cavities. It relied entirely on external beams for physics, because at these intensities, internal targets would create completely unmanageable radioactivity problems. The output would be either a short burst or a continuous beam. The proposal to abandon colliding beams and concentrate on high intensity was a response to the reviews of the previous proposal (see the next-to-last paragraph above). MURA also tried to do a better, less wishful job of cost estimation.

This was the proposal AEC wanted and they cheerfully sent it out for review. While the reviews were going on, AEC sent a cost expert, Phil McGee, who spent many weeks at MURA, checking but doing very little to change the original estimates of the cost of components. However, he introduced MURA to the wonders of EDIA (engineering, design, inspection, and administration), escalation and contingency and, following AEC guidelines on these, pushed the cost up to approximately US$ 100 million.

Revision of the 1962 proposal: A 12.5 GeV spiral sector accelerator was proposed to be built in seven years at a total cost of US$ 154,633,000, including lab and personnel space. The design intensity was 2×10^{14} protons per second.

None of these proposals was ever funded. The MURA people were disappointed, but in retrospect it is clear that the decision not to build a high-energy FFAG accelerator was a good one. The proposed MURA machines were generally well-designed, and would almost certainly have performed as promised in the proposals, but FFAG machines are very large and expensive. Smaller and less expensive ways to get high intensities and colliding beams were developed, some with contributions from MURA personnel (Secs. 3.7 and 4.7).

4.7. THE 1959 WORKSHOP: SYNCHROTRONS CATCH UP

One of the responses to the 1958 proposal was that MURA did not have enough contact with experimental work in particle physics. (Most of this section is taken from F. T. Cole's, Oh Camelot! [Cole, 1994, Sec. 17.1]) To try to correct this lack, MURA held a summer study in 1959 on the uses in particle physics of high-intensity beams from FFAG accelerators. MURA went to some length to prepare for the influx of people, but, as often happens, some of those invited went off on different tangents and, looking back, the seeds of MURA's demise were sown there.

Instead of thinking about FFAG, Matthew Sands (Fig. 4.6) of the California Institute of Technology proposed [Sands, 1959] building a 300 GeV proton synchrotron, the novel feature of which was that a smaller synchrotron was used to inject into the large ring. This idea had occurred to a number of people. In 1956, Lee Teng had proposed [Teng, 1956] using a cyclotron as the injector to a synchrotron and Wilson had mentioned the idea in his 1955 *Handbuch der Physik*

Fig. 4.6. Matthew Sands at the 1959 MURA Summer Study.

article [Wilson, 1959], although in a sentence so convoluted that most people thought he was saying that Salvini had invented the idea. Like all first proposals, Sands work went too far in claiming a small aperture and very small costs, but it was a powerful idea, not to be neglected. An acceleration scheme with a sequence of circular machines, the extracted beam from each being injected into the next, is often called a "cascade synchrotron." A crucial question is whether injection and extraction from and into circular machines can be accomplished in such a way as to preserve, or nearly so, the beam phase density. If so, then cascade synchrotrons can reach the high output beam intensities promised by FFAG synchrotrons and can reach these intensities at higher energies and at lower cost. There were other things done at the summer study, but compared with Sands' idea they paled into insignificance. Perhaps the most important part of Sands' proposal was that it made a much-higher-energy accelerator seem well within the bounds of possibility. Cascade synchrotrons have lived up to their promise. Virtually all high-energy accelerators are now designed this way.

It is also interesting as a sidelight that T. Kitagaki was present at this summer study. But he was interested in a new concept of his — the "scanning-field accelerator," a rival to FFAG — and he never looked back at his proposed separated-function geometry for large rings. The separated-function geometry was much used in electron–positron storage rings, but in 1966, Gordon Danby pointed out its advantages for very large rings. It would have been valuable had people realized these advantages sooner.

Sands went home to Caltech that fall and immediately formed a group to work further on the idea — the Western Accelerator Group (WAG). Ernest Courant, Hildred Blewett, Kenneth Robinson of the Cambridge Electron Accelerator and others contributed to this effort on visits there.

4.8. THE DIRECTORSHIP OF BERNARD WALDMAN

By late 1959, the laboratory was under great stress. The laboratory had claimed great things for FFAG accelerators in achieving colliding beams of high intensity, but the demonstration of these virtues in the 50 MeV model was not working out. The beam lifetimes were too short because of a small dynamic aperture. The largest part of the problem was that the magnetic fields in the "nonscaling pole" region did not produce zero (or small) chromaticity. The magnet power supply did not provide adequate regulation to achieve stable operation. In addition, the injection system, based on a commercial X-ray tube and a rudimentary transport system, could not deliver the proper emittance for efficient injection. In the early winter of 1960, attempts were still being made to stack enough beam to demonstrate the virtues of the two-way FFAG concept. The MURA Board of Directors sensed the frustration of this situation. H. R. Crane, University of Michigan, then President of the Board, paid several visits to MURA to discuss the situation. The Board's response was to bring in Bernard Waldman from Notre Dame as director of the laboratory in place of Rollefson.

At Notre Dame in the 1930s, Waldman had helped develop the Van de Graaff accelerator, and with George Collins had started a nuclear physics program. From his work at Los Alamos and then at Tinian during the Manhattan Project, Waldman had developed a respected reputation and established connections with many notable Los Alamos alumni. He was familiar to the MURA staff, having spent several summers at MURA.

With his appointment as director of the laboratory, Waldman took leave from Notre Dame. His first action was to interrupt the operation of the 50 MeV Model and to start an intense effort of measuring the magnetic fields and correcting the problems. This effort was carried out by mobilizing the entire laboratory and by remeasuring and correcting the magnetic fields by working on a 24-hour schedule. A new power supply improved the regulation by a factor of 10. A completely new electron gun and injector system was developed, immersed in the magnetic field and better matched to the acceptance of the "one-way" operating point. By 1963, all the goals of the model had been achieved. More than 10 A of electrons were stored, with the intensity determined by fundamental limits, and the lifetime determined by emittance growth due to synchrotron radiation.

Bernard Waldman (1913–1986)

Bernard Waldman was one of those pioneering physicists who introduced nuclear physics and accelerators into the Midwestern universities. World War II led him to make contributions at the national level in the Manhattan Project and elsewhere. At a critical moment he agreed to lead the MURA Laboratory to success with the Wisconsin Model and to develop proposals for the future of the laboratory.

Bernie joined Notre Dame University as an instructor in 1938, after receiving his doctorate. He started his research in experimental nuclear physics using an electron electrostatic generator. He often joked that he and his colleague George Collins had slowed the progress of physics for some months in the 1940s by reporting a slightly incorrect value for the binding energy of the deuteron, caused by a small calibration error in their voltage measurement. Bernie had severe scars on the back of his hands from radiation burns acquired during this time. In fact, on one occasion, he was given only a short time to live by some doctors.

During World War II, Bernie worked on the Manhattan Project with Norman Ramsey and Luis Alvarez. He was present on the mission that dropped the atomic bomb on Hiroshima. After the war he returned to Notre Dame, for which he had very deep feelings of loyalty, even though he was a Protestant. He continued his work on improving the electrostatic accelerator and doing research on photodisintegration thresholds of deuterium and beryllium. He participated in MURA workshops, and in 1960, he was called by Dick Crane, Chairman of the MURA Board of Directors, to become director of the laboratory. This was at a time when the laboratory was having difficulty achieving the expected performance of the 50 MeV Two-Way Model accelerator. He mobilized the laboratory to attain their goals of performance and then to spearhead the drive to achieve funding for the FFAG high-intensity accelerator proposal. Upon the refusal of the government to fund the MURA proposal, Bernie returned to Notre Dame in 1964 and became Dean of the College of Science. He retired from Notre Dame in 1979, but continued his active involvement with government panels and the National Science Foundation at Michigan State University until his death in 1986.

During the war, Bernie had been a navy engineering officer involved in the construction of bases in the continental US. He told with relish the story of his first assignment after the war, when he was sent back to the Point Mugu base in California, where he had installed all the utilities. It had been done at such white heat that there were no drawings and the staff were waiting eagerly to plumb the depths of his memory as to where he had put the water and sewer lines. He was still a reserve naval officer at the time we were in Madison, and once in a great while we would see him dressed for a meeting in his captain's uniform, glittering with gold braid. Like every true navy man, he had infinite faith in painting visible surfaces and the color was always gray. We lived in laboratories with gray walls and floors for all the MURA years.

Horace Richard Crane (1907–2007)

Horace Richard (Dick) Crane was, of all the physicists associated with MURA, the most comprehensive in his knowledge and the most impressive in the wide range of his contributions to physics. In addition, he was a delightful person.

Dick was born in Turlock, California, and received his B.S. and, in 1934, his Ph.D. (*cum laude*) from the California Institute of Technology. In 1934, he joined the University of Michigan's Physics Department, where he remained until his retirement in 1978; he served as department chair from 1964 to 1972. He invented the concept of straight sections in synchrotrons — the "race track synchrotron," a design concept incorporated in virtually every synchrotron built since 1950. His pioneering measurements on the gyromagnetic ratio of the free electron are a cornerstone of quantum electrodynamics. His analyses of helical structure in molecules continue to be significant in genetic research.

From the beginning of MURA in 1953, Dick was an active member of the community's senior physicists, both in the administration of MURA, and as a contributor to the MURA physics discussions, particularly during autumn 1954 and summer 1955, when the MURA technical activities were centered at Ann Arbor.

His activities were wide ranging. During World War II, he worked on radar at the Massachusetts Institute of Technology and on the proximity fuse at the Carnegie Institution of Washington. He served as the director of the proximity fuse research at the University of Michigan and as the director of the atomic research project for the Manhattan District. He was a consultant for the National Defense Research Commission and the Office of Scientific Research and Development. He was President of MURA from 1957 to 1960, and served on its board throughout the existence of MURA. He was President of the American Association of Physics Teachers in 1965, on the Board of Governors of the American Institute of Physics from 1964 to 1971, and its Chairman from 1971 to 1975.

Dick's awards included the Naval Ordnance Development Award (1945), the Distinguished Alumni Award from Caltech (1967), the Oersted Medal (1976), the Distinguished Service Award and the Henry Russell Lectureship at the University of Michigan (1967), a Citation for Distinguished Service by the American Association of Physics Teachers (1968), and the National Medal of Science, awarded by President Ronald Reagan (1986). He was a columnist for the journal *Physics Teacher*, writing articles on "how things work"; this became a book, and a bestseller for the American Institute of Physics. He developed many exhibits for the Ann Arbor "Hands-On Museum" during his retirement years.

Dick's many hobbies included amateur radio, raising orchids and cacti, playing the violin, and fishing. He was a devoted family man, long-time married to his wife, Florence. He particularly enjoyed family and friends. Simply the way he lived — a balanced scientific life — made him a role model to his colleagues.

MURA Directors

From the beginning of studies by the Midwest group (first called the "Midwest Accelerator Conference") in 1953, Donald W. Kerst acted as Technical Director of the group. The Midwestern Universities Research Corporation (MURA), was formed in the autumn of 1955. In 1956, MURA established a laboratory in Madison, Wisconsin, the site preferred by MURA to locate its future high-energy physics facility. It appointed P. Gerald Kruger as "Director of the Laboratories," in recognition of the ongoing work at member universities, and reaffirmed Kerst as Technical Director. Under Kerst and Kruger the laboratory "in the Nash Garage" on University Avenue was put into operation, consolidating the efforts of the Midwestern universities.

In May 1957, Kerst and Kruger resigned in response to a letter, received from Lewis Strauss, stating that AEC would not support a second laboratory in the Midwest and that the MURA activity must be incorporated into the Argonne program. After carefully polling the staff of the MURA Laboratory and associated university faculty, the MURA Corporation decided to proceed with operations aimed at exploiting the "Two-Way FFAG" invented by Tihiro Ohkawa. It would construct a 50 MeV electron model of this system to investigate construction techniques and unknown phenomena, and submit a proposal for a colliding beam facility. To carry out this work, Ragnar Rollefson and Keith Symon of the University of Wisconsin agreed to act as Director and Technical Director, respectively. A 1958 proposal for a 15 GeV p–p collider facility was withdrawn, because the cost was judged to be too high.

The first operation of the 50 MeV Model took place in December 1959. The results were disappointing, and the Corporation selected a new Director — one of the founders, Bernard Waldman of Notre Dame University. Francis T. Cole was appointed Associate Director. The high-energy physics world had changed. It now wanted high intensity, not high energy. The improvements in the 50 MeV Model reflected this change, moving away from the two-way operating point and emphasizing high-intensity operation. In 1962, the laboratory prepared a proposal for a 10 GeV high-intensity accelerator using the spiral sector technique. The results from the improved 50 MeV model were so good that the Ramsey Panel recommended that the energy be increased to 12.5 GeV, but that the MURA proposal should take second place to the 200 BeV Project at Berkeley. Waldman, Cole, and the Corporation under the leadership of its President, Elvis Starr, together with several senators of Midwestern states, waged a political campaign to secure funding for the project. In early 1964, the proposal was rejected by President Lyndon Johnson and not included in the budget.

In 1964, many staff, including Cole, began to leave as others pondered their future. Mills was appointed Associate Director. Responsibility for the operation of the laboratory was shifted from AEC to Argonne Laboratory, and its program included ZGS improvement, continuation of studies with the 50 MeV accelerator, programs for the benefit of the 200 BeV Project including linac studies and magnet studies, and studies of an ultrahigh-energy accelerator. In 1965, Waldman returned to Notre Dame as Dean of Science, and Mills was appointed Director. Mills' mission, now as a professor of the University of Wisconsin, was

(Continued)

MURA Directors (Continued)

to carry out the programs above and to facilitate the entry of the MURA Laboratory into the University of Wisconsin as the Physical Sciences Laboratory. He introduced the 240 MeV storage ring desired by the remaining staff into budget negotiations with Argonne and secured its funding. On July 1, 1967, the MURA Laboratory was no longer in existence, the 200 BeV programs were transferred to NAL (the new National Accelerator Laboratory, later known as Fermilab) together with their staff, and the Physical Sciences Laboratory began operation. Only one MURA staff member accepted Argonne's blanket job offer. He was laid off after a few years. Both the Synchrotron Radiation Center (the outgrowth of the 240 MeV storage ring) and the Physical Sciences Laboratory are, 40 years later, thriving scientific institutions, due to the excellence of the MURA scientific and technical staff, and the commitment of the University of Wisconsin to their success.

To do this, it was necessary to stabilize a transverse instability by feedback, sweep positive ions out of the beam, and correct the demagnetizing effect on the magnets due to the beam current. All of these efforts received strong support from Waldman.

The success of the Brookhaven AGS and the CERN PS in achieving proton beams of high energy and great intensity resulted in considerable discussion concerning the role of FFAG in providing beams for future physics experiments. Also, the 1959 MURA summer study had suggested that colliding beams at higher energies could more easily be achieved using intersecting storage rings. Could the MURA 10 GeV proton spiral sector FFAG proposal be defended against proponents of these other machines? It was Waldman's duty to make the necessary decisions and defend the laboratory's future course of action. He responded to the challenge and mobilized the laboratory to push ahead with the FFAG proposal. Of course, the gallant effort came to an end with the report of the GAC/PSAC panel on High Energy Accelerator Physics and the decision by President Lyndon Johnson not to continue the funding of the MURA Laboratory. Even then, Waldman supported the continuation of important programs. For example, the magnet development and the linac development programs continued with greater emphasis. He was not interested in those activities aimed at changing the mission and purpose of the laboratory.

Waldman returned to Notre Dame in 1965 to become Dean of Science, and Fred Mills was appointed director of the laboratory during the transition years as the mission of the laboratory turned to assisting ANL and later became the Physical Sciences Laboratory of the University of Wisconsin. Mills also accepted a position as Associate Professor of Physics at the University of Wisconsin. As Director of the MURA Laboratory, he was an *ex officio* member of the MURA Board of Directors and was appointed Treasurer of the MURA Corporation.

4.9. THE PANELS AND THEIR RECOMMENDATIONS

During the 1950s and 1960s, a number of panels were convened to advise various government agencies, particularly the Atomic Energy Commission, on matters relating to high-energy physics. Copies of their reports may be found in *High Energy Physics Program: Report on National Policy and Background Information of the Joint Committee on Atomic Energy, Congress of the United States, February, 1965* [Congress, 1965]. Many of these reports had an influence on the fortunes of MURA. For completeness we list here all the major recommendations of the various AEC Panels. The MURA response is described in the final section of this chapter.

Bacher Panel Report: Report of the National Science Foundation Advisory Panel on Ultrahigh Energy Nuclear Accelerators, May 2, 1954.

Members of the Panel: R. F. Bacher, California Institute of Technology, Chairman; S. K. Allison, University of Chicago; H. A. Bethe, Cornell University; L. J. Haworth, Brookhaven National Laboratory; W. K. H. Panofsky, Stanford University; I. I. Rabi, Columbia University; M. G. White, Princeton University; J. H. Williams, University of Minnesota; J. R. Zacharias, Massachusetts Institute of Technology.

Invited representatives: AEC — Drs. T. H. Johnson, G. A. Kolstad; DOD — Drs. E. R. Piore, E. Montroll, W. E. Wright; Brookhaven National Laboratory — Dr. J. P. Blewett; MIT — Dr. M. S. Livingston; University of Illinois — Dr. D. W. Kerst; Harvard University — Dr. N. E. Ramsey, Jr.; Cornell University — R. R. Wilson; University of Wisconsin — R. Rollefson; NSF — Drs. A. T. Waterman, P. E. Klopsteg, R. J. Seger, P. H. McMillen.

Recommendations of the Panel:

(1)   That the government, through AEC, DOD, NSF and other interested agencies, continue active support of high-energy physics, including the design, construction, and operation of high-energy accelerators and experimentation with these machines.

(2)   That adequate support be given to existing research programs in this field.

(3)   That the government agencies should be receptive to proposals by qualified groups, for the construction of high-energy machines.

(4) That policy permit initiation of additional machines without waiting for full completion of prior construction programs.

(5) That proper attention be given in the choice of further accelerators to the need for accurate quantitative data as well as to the need for exploratory experiments.

(6) That no fixed general policy be made with regard to the location of new accelerators at individual universities, national laboratories or other research establishments, but that each proposal in the future be reviewed on its merits with due regard to the research which will be done, the stimulation to science, and the opportunities for training.

Haworth Panel Report: Report of the Advisory Panel on High Energy Accelerators to the National Science Foundation, October 25, 1956.

Members of the Panel: L. J. Haworth, Brookhaven National Laboratory, Chairman; S. K. Allison, University of Chicago; E. J. Lofgren, University of California, Berkeley; W. K. H. Panofsky, Stanford University; I. I. Rabi, Columbia University; A. Roberts, University of Rochester; F. Seitz, University of Illinois; R. Serber, Columbia University; M. G. White, Princeton University; J. H. Williams, University of Minnesota; J. R. Zacharias, Massachusetts Institute of Technology.

After commenting on the significance, cost, and diversity of support of high-energy physics, the Panel made the following recommendations:

(1) That the government continue active support of high-energy physics, including the design, construction, and operation of and experimentation with high-energy accelerators.

(2) That DOD, AEC, and NSF each engage directly in the support of high-energy physics; in particular, that DOD and NSF extend their support in this field to maintain important positions.

(3) That adequate support be given to existing research programs in this field.

(4) That planning for the support of high-energy accelerators anticipate an annual rate of expenditure in this field of US$ 60–90 million per year by 1962.

(5) That need for accelerators of a variety of characteristics be recognized. The most important parameters are energy, intensity, and kind of particle. In situations where a choice is to be made, it is usually better to extend the range of these important parameters than to increase the number of functionally similar accelerators.

(6) That adequate support be given to research and development pointed toward new types of accelerators. It is especially recommended that efforts be made to increase the energy limit.

(7)   That no fixed general policy be made with regard to the location of new accelerators at individual universities, national laboratories, or other research establishments, but that each proposal be reviewed on its merits with due regard to the research that will be done, the stimulation to science, and the opportunities for training.

The Panel made various, more extensive comments on these recommendations, including a comment on the need for an accelerator in the Midwest. In particular, in an appendix on future lines of development, the Panel discussed the FFAG principle developed by MURA, which promises high-intensity beams and the possibility of experimenting with colliding beams to achieve much higher center-of-mass energies.

Supplemental Report of the Haworth Panel: Supplement to the Report of the National Science Foundation Advisory Panel on Ultrahigh Energy Nuclear Accelerators, August 7–8, 1958.

Members of the Panel in August 1958: L. J. Haworth, Brookhaven National Laboratory, Chairman; H. L. Anderson, University of Chicago; H. R. Crane, University of Michigan; B. T. Feld, Massachusetts Institute of Technology; E. J. Lofgren, University of California, Berkeley; L. I. Schiff, Stanford University; F. Seitz, University of Illinois; R. Serber, Columbia University; M. G. White, Princeton University; R. R. Wilson, Cornell University.

The report summarizes the advances made in the two years since the 1956 report, including models and studies of FFAG at MURA, T. Ohkawa's two-way FFAG machine, and the development of storage rings as an alternative way of achieving colliding beams. It then states that none of these developments changes the recommendations of the 1956 report, and lists a series of recommendations which are almost word for word those of the 1956 report. It emphasizes the importance of the search for new methods of guiding and accelerating particles. It states that many of the important accelerator advances of recent years have come from the MURA group, and urges that the MURA group be supported on a continuing basis with funds and facilities necessary for its participating intensively in the development, construction, and operation of accelerators.

Piore Panel Report, 1958: US Policy and Actions in High Energy Accelerator Physics. Report by a Special Panel of the President's Science Advisory Committee and the General Advisory Committee to the Atomic Energy Commission.

Members of the Panel: SAC — Dr. Emanuel Piore (Chairman), Dr. Hans A. Bethe, Dr. L. J. Haworth; GAC — Prof. Jesse W. Beams, Dr. Edwin M. McMillan.

After an extensive discussion on the status and support of high-energy physics and particle accelerators, the Panel recommended that the federal government:

(1) Expand its support of high-energy accelerator physics.

(2) Plan for an increasing level of support for construction and operation of high-energy accelerators to a level of approximately US$ 125 million by fiscal year 1963.

(3) Note that the increased funds for high-energy accelerator physics should be provided without affecting the support of other areas of basic science.

(4) Note that specific needs exist for an electron linear accelerator of at least 10 BeV and a high-intensity proton accelerator of at least 8 BeV.

(5) Approve the initiation in fiscal year 1959 of the proposed accelerator project at Stanford University (electron linac of at least 10 BeV) ... (particular relevance to DOD).

(6) Note that construction of the Stanford accelerator can be expected to cost approximately US$ 100 million.

(7) Note that the initial operating costs of the Stanford accelerator will be approximately US$ 15 million per year once it comes into operation and that instrumentation and initial experimentation should be undertaken several years in advance of completion of the accelerator.

(8) Reject the present MURA accelerator proposal but continue to provide adequate support and encouragement to the MURA development group.

(9) Reject the present Oak Ridge accelerator proposal but encourage the further exploration of the research needs of the region and the design of accelerators capable of meeting those needs.

(10) Note that the interests of the federal government and the scientific community are best served by the direct financing of projects in this field by AEC, DOD, and NSF, with the contract authority and budget responsibility for a given project being vested in a single agency.

(11) Approve the establishment of an interagency council at the policy level to assure coordination of budget and technical planning, this council to be assisted by the scientific staff of the agencies concerned.

(12) Approve the provision of contract funds for periods of between one and three years for the support of research operations connected with high-energy accelerators.

(13) Encourage international collaboration and cooperation in the planning for and design of future machines and the increased use of facilities.

(14) Request the National Academy of Science to study and advise the government on, the best method for proceeding with international cooperative

research on new accelerator concepts, such cooperative activity including the Soviet Union.

In the preliminary discussion, it is stated that there is no clear need for an extension of the energy parameter for protons beyond 30 BeV — a rather surprising statement. At that time, there were no known particles whose production would have required higher energies, and there was talk of an "asymptopia" beyond about 30 BeV where cross sections would approach asymptotic values and experimental data would not be very interesting. Politics probably also played a role, since higher energies were being proposed by MURA using colliding beams, and those from other laboratories would have tended to favor arguments against the interest in higher-energy experiments.

Piore Panel Report, December 15, 1960: Report of the PSAC-GAC Panel on High Energy Accelerator Physics, December 14, 1960.

Members of the Panel: Dr. E. R. Piore, International Business Machines Corp., Chairman; Dr. Leland J. Haworth, Brookhaven National Laboratory; Dr. Jesse W. Beams, University of Virginia; Dr. Hans A. Bethe, Cornell University; Dr. Edwin M. McMillan, Lawrence Radiation Laboratory, University of California; Dr. Eugene P. Wigner, Princeton University; Mr. George D. Lukes, technical assistant to the Panel.

Special consultants to the Panel: Dr. Robert F. Bacher, California Institute of Technology; Dr. Charles E. Falk, Brookhaven National Laboratory; Dr. M. Stanley Livingston, Cambridge Electron Accelerator, Harvard – Massachusetts Institute of Technology; Dr. W. K. H. Panofsky, Stanford University; Dr. Keith R. Symon, Midwestern Universities Research Association, University of Wisconsin; Dr. John H. Williams, University of Minnesota.

Invited participants: Dr. E. J. Lofgren, Lawrence Radiation Laboratory, University of California; Dr. G. K. O'Neill, Princeton University; Dr. R. R. Wilson, Cornell University.

Representatives of federal agencies: Mr. William A. Ellis, National Science Foundation; Dr. Jerome Fregeau, Office of Naval Research; Dr. John T. Holloway, Department of Defense; Dr. H. W. Koch, National Bureau of Standards; Dr. George A. Kolstad, Atomic Energy Commission; Dr. Paul W. McDaniel, Atomic Energy Commission; Dr. J. Howard McMillen, National Science Foundation; Dr. Randall M. Robertson, National Science Foundation; Dr. William E. Wright, Office of Naval Research.

Report of the Panel:

After a discussion on the background, including a reaffirmation of the conclusions of the panel's 1958 report, the report reviewed recent developments, including, among others, the fact that the US continues to play the leading role in high-energy physics research, the fact that the AGS is operating, and the increased recognition and support of high-energy physics abroad. After extensive discussion, the panel recommended that the federal government:

(1) Continue to expand its support of high-energy accelerator physics.

(2) Continue to support accelerators now in operation or under construction.

(3) Provide strong support for the development of new techniques of particle detection, data reduction, and data analysis.

(4) Increase the support of university high-energy research, including buildings and major equipment.

(5) Authorize construction of the Stanford accelerator at the earliest possible date.

(6) Plan for an increasing level of support of the program based on existing and authorized accelerators, including Stanford, that will reach US $170 million annually by 1965 and perhaps 10% more by 1970.

(7) Note the need for an additional program of construction to extend the energy and intensity parameters.

(8) Support design studies leading to accelerators of substantially higher energy.

(9) Support further detailed studies of the most effective means of achieving high-intensity beams of a wide range of secondary particles.

(10) Note the specific need for and the present feasibility of a very-high-intensity proton accelerator of 500–1000 MeV to serve as a high-intensity source for low-energy pions.

(11) Authorize the construction of relatively small accelerators at university sites when this can be justified.

(12) Note that progress in the field may justify future annual expenditure for construction and operation of new accelerators that, by 1970, may approach cost levels similar to those (US$ 175–200 million annually) expected to be needed by that time to support the present family of accelerators (recommendation 6). The total support would then be US$ 350–400 million annually.

(13) Continue to support advanced concepts of accelerator technology.

(14) Note that if all the needs cannot be met a proper balance should be maintained between support of present programs and extensions through new projects.

(15) Note that the ten-year forecast of the Interagency Staff Report is a well-considered document.

(16) Provide for a review of the high-energy accelerator program by a scientific panel whenever the situation warrants it, and in any case at no more than two-year intervals.

(17) Note that the interests of the federal government and the scientific community will be best served by maintaining a significant diversity of support for high-energy physics between the Atomic Energy Commission, the Department of Defense, and the National Science Foundation.

(18) Encourage international cooperation in the field of accelerator development and the mutual use of existing facilities.

(19) Note that the construction of an ultrahigh-energy accelerator through the mechanism of an international project involves many problems which should be thoroughly studied before the possible initiation of such a project. These studies should eventually become multilateral in character.

There is an addendum to the report, written by Eugene Wigner, in which he says that he supports the recommendations of the Panel, but is concerned that the report may have oversold its proposed program. He points out that, except for clarifying the weak interactions leading to beta decay in nuclear physics, high-energy physics has not had any practical applications to other fields. He feels that the report does not pay enough attention to the fiscal and manpower consequences of the proposed program for other fields. He suggests that more emphasis might have been given to clashing beams as a means of reaching higher energies. He points out the inconsistency in the Panel recommendations for both maintaining the US leadership in high-energy physics and promoting international cooperation, and suggests that there be fuller cooperation with CERN.

Ramsey Panel Report, May 10, 1963: Report of the Panel on High Energy Accelerator Physics of the General Advisory Committee to the Atomic Energy Commission and the President's Science Advisory Committee.

Members of the Panel: Dr. Norman F. Ramsey, Harvard University, Chairman; Philip H. Abelson, Carnegie Institution of Washington; Owen Chamberlain, University of California; Murray Gell-Mann, California Institute of Technology; E. L. Goldwasser, University of Illinois; T. D. Lee, Columbia University; W. K. H. Panofsky, Stanford University; E. M. Purcell, Harvard University; Frederick Seitz, National Academy of Sciences; John H. Williams, University of Minnesota.

Ex officio members: Randall M. Robertson, NSF (representing the Technical Committee on High Energy Physics of the Federal Council on Science and Technology); David Z. Robinson, Office of Science and Technology.

Executive Secretary: Johannes C. Severiens, AEC.

Participants: Air Force Office of Scientific Research — J. E. Duval, A. W. Harrison Jr., I. A. Wood; Argonne National Laboratory — A. V. Crewe, R. H. Hildebrand; AEC — L. J. Haworth (Commissioner), S. G. English, G. M. Kavanagh, G. A. Kolstad, L. J. Laslett, P. W. McDaniel, R. P. McGee, D. R. Miller, W. A. Wallenmeyer; Brookhaven National Laboratory — M. Goldhaber (Director),

J. P. Blewett, C. E. Falk, G. K. Green, G. F. Tape, L. C. L. Yuan; Bureau of the Budget — F. C. Schuldt, Stanley Small; Columbia University — Melvin Schwartz, Robert Serber; CERN — V. F. Weisskopf; University of California, Los Angeles — J. R. Richardson, B. T. Wright; Cornell University — D. A. Edwards, R. R. Wilson; Department of Defense — F. J. Weyl; Lawrence Radiation Laboratory — E. M. McMillan (Director), G. F. Chew, Denis Keefe, E. J. Lofgren, Lloyd Smith, G. H. Trilling; Los Alamos Scientific Laboratory — C. L. Critchfield, Louis Rosen; University of Michigan — L. W. Jones; Midwestern Universities Research Association — Bernard Waldman (Director), F. T. Cole, Aaron Galonsky, K. R. Symon; National Aeronautics and Space Administration — Harry Harrison; National Science Foundation — R. H. Bolt, Wayne Gruner, W. L. Kolthun, J. H. McMillen; Office of Naval Research — J. H. Fregeau, S. H. Krasner, W. E. Wright; Oak Ridge National Laboratory — R. S. Livingston, A. H. Snell, Alexander Zucker; University of Pennsylvania — Henry Primakoff; Princeton University — G. K. O'Neill; Stanford University — W. M. Fairbank, Robert Hofstadter, P. B. Wilson; University of Washington — R. W. Williams; Department of State — Ragnar Rollefson; University of Wisconsin — M. L. Good, R. G. Sachs; Yale University — V. W. Hughes, G. W. Wheeler.

After extensive discussion, the Panel report recommended that the federal government:

(1) Authorize at the earliest possible date the construction by Lawrence Radiation Laboratory of a high-energy proton accelerator at approximately 200 BeV energy.

(2) Authorize the construction of storage rings at Brookhaven National Laboratory after a suitable study.

(3) Support intensive design studies at Brookhaven National Laboratory of a national accelerator in the range of 600–1000 BeV. A request for authorization may be anticipated in about 5 or 6 years.

(4) Authorize in fiscal year 1965 the construction by MURA of a supercurrent accelerator without permitting this to delay the steps toward higher energy. The energy of the MURA FFAG accelerator should be 12.5 BeV instead of the 10 BeV originally proposed.

(5) Support the construction of the proposed 10 BeV Cornell electron accelerator, including plans leading to its evolution into a nationally available facility.

(6) Support the development and construction of electron–positron storage rings.

(7) Provide strong support for the development and utilization of new techniques of particle detection, data reduction, and data analysis.

(8) Increase the support of university high-energy users' groups for buildings, major equipment, and computational facilities.

William Anton Wallenmeyer (1926–)

William Anton Wallenmeyer joined MURA as a graduate research assistant while completing his Ph.D. During his years at MURA, he was involved, as an experimentalist, with a number of different projects.

Bill was born in Evansville, Indiana. He married Diane May Hankins and they have 4 children and 13 grandchildren. He was educated in Evansville, including 2 years at Evansville College, after spending 27 months in the US Army Air Corps. In 1948, he transferred to Purdue University, where he received a B.S. with distinction in 1950, an M.S. (Physics), and a Ph.D. (Physics) and a D.Sc. (Hon.) in 1989. His Ph.D. thesis was entitled "Meson Production in n–p Collisions at Cosmotron Energies." He published this as *Phys. Rev.* 105 (1957), republished by the Physical Society of Japan in *Selected Papers in the Field of High Energy Physics in the BeV Region.*

Bill was the first Purdue graduate student to be awarded his Ph.D. based upon research carried out at a national laboratory, in his case Brookhaven National Laboratory. In the autumn 1956, joined the MURA Laboratory, where, at first, he helped Haxby, Mills, Peterson, Day, and Radmer to complete the construction of the Spiral Sector Model. Upon the successful operation of that model, he made the measurements for the resonance and intensity survey. He then worked on the 50 MeV Model, including work on injection and magnets. In 1960, he together with Young and Pruett, led the effort to correct the fields in the nonscaling pole region. This was done purely by measurements, because believable three-dimensional field calculations were still three decades into the future. He managed the first months of successful operation in 1961–1962.

In June 1962, Bill joined Laslett at the AEC Division of Research in Germantown. Bill remained at AEC, then ERDA, then DOE, and was Director of the High Energy Physics Section or its successors for 24 years. He believed strongly that the program required input from those carrying it out, and one of his major innovations was the formation of the High Energy Physics Advisory Panel (HEPAP), which survives as the principal link between researchers in the field and the funding agencies for the field. He played a major role in the planning and funding of all aspects of the field for 25 years.

Bill retired from DOE in 1987, and became President of SURA (Southeastern Universities Research Association), which built CEBAF (Continuous Electron Beam Accelerator Facility) Laboratory in Newport News, Virginia. He joined URA (Universities Research Association, former manager of Fermilab) as a part time special assistant to the President of URA, at his request, from July 1992 to May 1993. In 1989, he was awarded an honorary degree by Purdue University. At the present time Bill is working on projects for the American Physical Society.

Marshall W. Keith (1912–1996)

Marshall W. Keith was the skilled administrative engineer who made the new MURA laboratory run smoothly. His experience in organizations and government operation aided immeasurably in the creation and functioning of the organization.

Marshal was born and raised in Crandon, Wisconsin, and was trained as an engineer at the University of Wisconsin. During World War II, he served as an officer in the US Navy. After the war, as a reserve officer, he served in the Office of Naval Research as a contracts officer managing research grants at the Midwestern universities. Since the Office of Naval Research funded most of the university research in nuclear physics in that period, he was well known by the nuclear physicists at the Midwestern universities, who formed MURA.

When the MURA Laboratory was formed, Marshall was a key figure in its formation and operation. He acted as business manager, head of engineering, personnel, shops, etc. He conceived the plan for MURA to purchase the IBM 704 computer as a way to save money for research. This role continued until the transition of the laboratory to the University of Wisconsin. He had hoped to play a similar role at NAL, but R. R. Wilson rejected that role for him. Following that experience he moved to the University of Minnesota. Eventually he retired and moved back to Crandon.

Marshall initiated many of the "stitches" that held the laboratory together. As an outstanding example, the fishing trips were his idea, and were held near his childhood home. He also organized some deer hunting trips in the same area.

(9) Close down or reduce the level of operation of accelerators which become relatively unproductive.

(10) Support the study of new accelerator principles and techniques.

(11) Recognize the need for adequate visitor housing (both short- and long-term) at the above recommended national facilities.

(12) Provide for a review of the high-energy physics program at suitable intervals.

Tables and figures are given for anticipated costs and manpower required by the recommended program.

On December 11, 1963, the Panel issued a supplementary statement in response to a request for clarification, stating that in view of various technical developments since April 1963, the Panel now believed that the originally proposed

William R. Winter (1930–1994)

William R. Winter was the mechanical engineer who put into reality the ideas of the physicists. He became an expert in stress analysis, strength of materials, welding, and most particularly vacuum system fabrication and processing. He managed the construction of most projects at MURA, and designed the building addition, which housed the electron storage ring.

Bill was raised in Milwaukee, Wisconsin, and graduated in Mechanical Engineering from the University of Wisconsin in 1953. After serving in the US Army from August 1953 to July 1955, he entered law school at UW. Planning to continue his education, he took a "temporary" job at MURA shortly after the laboratory opened its doors in the old Nash garage in September 1956. He was subjected to Marshall Keith's test: a truck with a particularly nasty load pulled into the garage, and Marshall told him to go and help to unload it. He passed that test, and all others, becoming chief engineer at MURA, then PSL and SRC.

Bill really came into his own with his work on the 50 MeV model vacuum chamber and magnets. He was instrumental in the development of bubble chambers for use at the ZGS at Argonne. He helped design and build the 30″ liquid hydrogen chamber, a workhorse of the ZGS program, later the model for the large heavy liquid bubble chamber. That chamber was never built at Argonne, but a twin chamber, "Gargamelle," was built at CERN and discovered neutral currents, which had been the goal of the US physicists who planned the Argonne chamber. Bill was also responsible for the engineering design of the Michigan–MURA cosmic ray experiments. Later, in addition to the University of Wisconsin Toroidal Multipole Plasma Physics experiment, he performed the engineering on the UW storage ring (Tantalus) and later the ring Aladdin. He built the successful magnet and vacuum models for NAL in 1967.

Each year, Bill would vanish for a week or so, to sail in the Mackinac Island Race from Chicago to Mackinac Island. His love for sailing stayed with him throughout his life, as did his love for engineering. Bill had an active hobby of model railroading. His basement was filled with tracks, stations, and other features. He was a loving husband to his wife, Barbara, and a devoted father to his children. They spent most of their vacation time touring the US and Canada in their recreational vehicle.

energy of 10 BeV for the MURA machine was more suitable, and that it should be located in the Middle West but that the Panel would not fix the exact site.

Later panels, not included in the 1965 High Energy Physics Program report [Congress, 1965], include [Cole, 1994, p. 34]:

The Good Panel (see also [Congress, 1965]), chaired by Myron Good, Stony Brook, included no directors or executive personnel of laboratories or agencies and was set up to provide a different viewpoint on the high-energy physics program. Its accelerator consultants were Francis T. Cole, MURA; John Blewett, Brookhaven National Laboratory; and Lloyd Smith, University of California, Berkeley. After considerable discussion, the Panel ended up endorsing the Ramsey Panel recommendations.

The Laslett Panel, chaired by L. Jackson Laslett, director of high-energy physics research at AEC, was formed to review the space charge limits for the ZGS, proposed by Argonne National Laboratory, and 12.5 GeV FFAG accelerator, proposed by MURA. Frances T. Cole represented MURA on the Panel, and Lee Teng represented Argonne. The panel concluded that the MURA proposal would provide a higher-intensity 12.5 GeV beam of protons.

4.10. MURA RESPONDS

Following the recommendations of the Ramsey Panel, MURA had chosen not to confront the Lawrence Radiation Laboratory on the high-energy frontier, but to pursue the higher intensities that were possible with FFAG accelerators. This led to a proposal for a 12.5 GeV spiral sector accelerator with a 200 MeV proton linac as injector.

All aspects of this device were being modeled, and prototypes were under construction. On the political side, Elvis Starr (then President of Indiana University), the MURA Board of Directors, and Benard Waldman (Director of MURA) mounted a campaign involving Midwestern Governors, Congressmen and Senators, most notably Hubert Humphrey of Minnesota and Gaylord Nelson of Wisconsin (the other Wisconsin Senator, William Proxmire, an avid MURA supporter, demurred because he was at odds with Lyndon Johnson), to move the project through the balky federal bureaucracy and Congress. The group believed that President Kennedy would include the project in the next budget under preparation in the autumn of 1963.

CHAPTER 5

THE LAST YEARS OF MURA, 1963–1967

5.1. THE END OF MURA

On November 22, 1963, President John F. Kennedy was assassinated. At MURA, this was announced to the staff and some of them gathered in the hall to discuss it. There was a sense of foreboding that the change in the administration would lead, yet again, to the frustration of their efforts.

After Kennedy was assassinated, the Midwestern political drive made it all the way to the White House, where President Johnson met with the Midwestern people early in 1964. It was clear to many of those present that he had made up his mind before the meeting to turn it down. The reason given was that he needed to keep the budget below US$ 100 billion. (This was the first year that the social security funds were included as a line item within the federal budget, rather than as a separate item.)

Following this political action, in early 1964, the MURA scientific and engineering staff held a series of seminars to discuss their possible futures. At issue were both the future, if any, of the laboratory and the prospects of the individuals. As things developed, there was considerable choice available to individuals. First, Argonne National Laboratory offered to take the entire staff and move them to their site at Lemont, Illinois. Second, the Midwestern Universities would offer jobs to the scientists in departments at their campuses. Third, the University of Wisconsin would take the administrative, technical, engineering, and some scientific staff to operate the laboratory in support of its many research programs. Fourth, AEC agreed to continue to support programs that were important to future projects, particularly the 200 BeV Project. These programs would end when the 200 BeV Project began construction. As part of the closeout costs, the lab was to offer help to the ZGS at Argonne, which was beginning to operate.

In these seminars, a number of matters were worked out. Many individuals left to join other laboratories or universities (Cole, Christian, Meads, Galonsky, Swenson). Others continued work on projects important to the 200 BeV Project

The Board of Directors of the Midwestern Universities Research Association

Below is a list of the members of the MURA Board over the years 1956–1967, grouped according to the universities they represented.

University of Chicago
Dr. R. H. Hildebrand
Mr. W. B. Harrell
Dr. G. W. Beadle

University of Iowa
Dr. V. M. Hancher*
Dr. J. A. Jacobs
Mr. E. T. Jolliffe
Dr. M. Dresden
Dr. H. Bowen

University of Michigan
Dr. H. R. Crane
Mr. W. K. Pierpont
Dr. H. H. Hatcher*

Northwestern University
Dr. J. R. Miller*
Dr. R. Fisher
Dr. M. Dresden
Mr. A. T. Schmehling
Dr. M. M. Block
Mr. W. S. Kerr

Purdue University
Dr. F. L. Hovde*
Mr. R. B. Stewart
Dr. T. R. Palfrey
Mr. L. J. Freehafer

University of Illinois
Dr. P. G. Kruger
Mr. H. O. Farber
Dr. D. D. Henry*
Dr. E. Goldwasser

Iowa State University
Dr. D. J. Zaffarano
Dr. J. H. Hilton
Mr. B. H. Platt
Dr. L. J. Laslett
Dr. C. Hammer

Michigan State University
Dr. J. A. Hannah*
Mr. P. J. May
Dr. H. Blosser
Dr. A. Galonsky

Notre Dame University
Dr. T. M. Hesburgh*
Dr. B. Waldman
Mr. G. E. Harwood
Dr. C. J. Mullin

Washington University
Dr. R. Sard
Dr. E. U. Condon
Dr. C. Tolman
Mr. J. H. Ernest
Dr. F. B. Shull
Dr. T. H. Eliot*
Dr. G. W. Hazzard

Indiana University
D. A. C. G. Mitchell
Dr. E. J. Stahr*
Mr. J. A. Franklin
Dr. H. B. Wells
Dr. H. Bowen

University of Kansas
Dr. M. Dresden
Dr. J. D. Stranathan
Dr. W. C. Wescoe*
Mr. R. Nichols

University of Minnesota
Dr. J. H. Williams
Mr. L. R. Lunden
Dr. A. O. C. Nier
Dr. O. M. Wilson

Ohio State University
Dr. H. H. Nielson
Dr. N. G. Fawcett*
Mr. G. B. Carson

University of Wisconsin
Dr. R. Rollefson
Mr. A. W. Peterson
Dr. C. A. Elvehjem*
Dr. F. H. Harrington*
Dr. W. F. Fry

* University presidents

until construction began in 1967 (Young, Curtis, Owen, Snowdon). Snowdon and Young accepted professorships in the UW Nuclear Engineering Department to carry out a program in accelerator physics in case it could obtain funding.

On the physics frontier, Mills proposed construction of a small (240 MeV) electron storage ring using the 50 MeV accelerator, with its extracted beam, as injector. It would be employed to study the accumulation of intense beams, in part by using synchrotron radiation as an emittance damper, a scheme he had proposed in 1962 for ACO, the Orsay storage ring. Later, Gerry Kruger communicated to the group, through Ed Rowe, the report of a National Academy of Science panel that described the need for such a ring as a source of ultraviolet radiation for research in many fields. Mills, Rowe, Pruett, and Winter decided to pursue the storage ring, first for accelerator physics, and then, if possible, for synchrotron radiation research. (When the Physical Sciences Laboratory tried to obtain AEC funding for this program it was told, remarkably, by Bill Wallenmeyer, then at AEC, that such a program was not needed because the 200 BeV Project was the last accelerator that would ever be built.)

Within a year, others proposed using MURA, or its successor, the Physical Sciences Laboratory (PSL), to manage or build their research projects. Kerst, now at UW Madison, proposed construction at the laboratory of a large toroidal multipole plasma containment device for fusion research. Jones at Michigan proposed using the MURA Laboratory to help mount a large mountaintop cosmic ray experiment to get an early peek at the physics to be done in the 200 BeV Project.

When Cole left in 1964, Mills was appointed Deputy Director of MURA. AEC decided that it would no longer speak directly to MURA, but would only communicate through Argonne. When Bernard Waldman returned to the University of Notre Dame as Dean of Science, Mills accepted the job as Director. He also accepted an associate professorship at the University of Wisconsin, and was expected to be the Director of the PSL, which would be part of the UW Graduate School. For the period January 1964 – June 1967, the program of PSL included the ZGS Tuneup and Improvement Program, the 200 MeV Linac Program, Magnet Development, the Cosmic Ray Project, and the Electron Storage Ring Project. The last two of these activities continued under PSL, while the first three terminated June 30, 1967. These activities are described in more detail in the next sections.

5.2. THE ZGS TUNEUP AND IMPROVEMENT PROGRAM

During 1955, a few senior American physicists were invited to visit the Soviet Union and Dubna Laboratory (north of Moscow). It was there they saw the Russian

Donald A. Swenson (1932–)

Donald A. Swenson is a skilled experimental physicist who made immediate contributions to the MURA program in the areas of injection, beam dynamics, instabilities, and linear accelerator design. He demonstrated experimentally the effect of "Landau damping" before the effect had been theoretically defined. He led the design and construction of the 100 MeV injector linac for the Los Alamos Meson Physics Facility (LAMPF) project at Los Alamos. He invented the "post-couplers" for Alvarez linacs, which provide strong stability against frequency errors.

Don joined the MURA staff after completing his Ph.D. in Physics at the University of Minnesota in 1958. His first contributions were made in the development of beam diagnostic techniques and he was able to observe the negative-mass instability. When the program for the development of the 200 MeV linac was initiated, he contributed by developing the beam dynamics computer program called PARMILA (Phase And Radial Motion In Linear Accelerators), which continues to be the leading beam dynamics program for linacs.

Don has been active in the particle accelerator field throughout his career. He was employed at Los Alamos National Laboratory from 1964 to 1983. There he played a leading role in the design and development of the drift-tube linac, invented a method of rf field stabilization using post-couplers, was the Principal Investigator for the PIGMI Program (Pion Ion Generator for Medical Irradiation), and designed and built the first RFQ (Radio Frequency Quadrupole) linac in the Western world. From 1983 to 1986, Don was a Visiting Professor at Texas A & M University, where he continued to develop and improve computer programs useful in accelerator design. During the "Star Wars" years, he worked at Science Applications International Corporation (SAIC) from 1986 until 1991, where he continued to develop the RFQ for more practical applications, designed and built a new linac structure called the "disk-and-washer" structure, and proposed a practical fabrication scheme for a superconducting RFQ linac. He joined the Superconducting Super Collider Laboratory in 1991. He is a founder and the Chief Scientist of Linac Systems at Los Alamos (1991–), a company whose mission is to develop compact, reliable, and inexpensive proton and ion linacs for scientific, medical, industrial, defense, and homeland security applications. Don is the inventor of the very efficient and compact RFI (Rf Focused Interdigital) linac structure.

Don is an extremely talented experimental accelerator physicist with the ability to incorporate advanced computer programs into his design efforts. He is recognized for his dogged and single-minded pursuit of an idea that others may have cast aside. New ideas and concepts abound in his professional and personal life. Currently he has over 50 years of practical experience with particle beam systems and over 15 years as Principal Investigator of accelerator-based projects. He holds 17 patents for accelerator and beam diagnostic systems. Don's wife, Barbara, has declared that she is thankful for Don's present involvement in Linac Systems, because it provides an outlet for his creative talents.

Cyril D. Curtis (1920–)

Cyril D. Curtis played an important role in all activities related to beam injection. Most of all, he is noted for his development at MURA of ion sources and the "short" or "high field" accelerating column for the Cockroft–Walton preinjector, put into use at Fermilab and elsewhere.

During World War II, Cy became a Radar-Weather Officer in the US Army Air Force. His experience resulted in his being sent to Los Alamos Laboratory to assist in testing the first atomic bomb at the Trinity test site. He witnessed the explosion from a bunker six miles from ground zero. After the war, he received his Ph.D. degree from the University of Illinois, doing his thesis with the use of Don Kerst's 80 MeV betatron.

In 1951, Cy joined the Naval Reactor Division of Argonne National Laboratory, where he engaged in experiments with the Zero-Power nuclear reactor, establishing the groundwork for the nuclear power plant of the submarine *Nautilus*. He also participated in the first experiment to demonstrate, in a fast neutron reactor, the production of more plutonium (from uranium-238) than the consumption of nuclear fuel, i.e., the breeder concept. From 1953 to 1959, he taught physics at Vanderbilt University and directed thesis students in electronic instrumentation and low-energy nuclear physics.

In 1959, Cy joined MURA. Among his various activities were design work on an electron gun for the 50 MeV Two-Way Model and many experiments with the accelerator, including the study of various multiturn injection schemes, beam instabilities, and simultaneous acceleration of two electron beams in opposite directions. Also, work began on duoplasmatron ion sources and accelerator column development in anticipation of a proton accelerator. This work continued at Fermilab, which Cy joined in 1967, and resulted in an operating preaccelerator for the 200 MeV linear accelerator. Later, a negative ion (H⁻) source was added to aid multiturn injection into the Booster synchrotron. Cy played a significant role in the commissioning of the linac at Fermilab and aided in the commissioning of the total accelerator complex. He also had a role in establishing the neutron therapy program at Fermilab and in developing the proton synchrotron for cancer treatment for the Loma Linda University Medical Center in California. He served as a consultant to LLUMC from 1984 to 1990. Cy had a penchant for creating new technical devices (from neutron detectors to voltage generators), prompting Fred Mills to christen them "Ψ-machines."

Fred Mills comments: "Once in those final MURA years, when Cy was developing his beautiful ion source and preaccelerator (the one guaranteed stop on a tour of Fermilab), Cy came to me and wanted to build a small device which appeared to me to be a rotating variable capacitor similar to that used to tune AM radios. I doubted its functionality, but told him to go ahead. Only after I retired did I have time to catch up on QED and the quantum vacuum. Something I read led me to think about how one would interact with the zero point energy (the Casimir force). It turned out to be Cy's machine. In spite of his shy retiring manner, he was always 40 or 50 years ahead of us all."

In retirement Cy organizes music recitals, plays the piano, and keeps his mind active by examining challenges to the second law of thermodynamics.

Ednor M. Rowe (1927–1996)

Ednor M. Rowe came to MURA from Purdue in 1956, having helped Bob Haxby build and measure the Michigan Model magnets. He quickly discovered his niche and, working with Larry Johnston of Minnesota, he established himself as the expert on radio frequency systems. In 1964, he joined MURA and led the effort to build an electron storage ring. In 1967, he stayed at the Physical Sciences Laboratory (PSL), and led the small team (Rowe, Pruett, and Winter), who, in addition to their other duties, would complete the construction of the storage ring. He managed the experimental program with the ring and, following its success, pursued the design and construction of a larger ring, called Aladdin.

Ed, a master electronic technician, was trained at Purdue University. He earned his M.S. in Physics in 1957. While at Purdue, he spent some time working on MURA projects and, after receiving his degree, he joined the MURA staff. In 1990, he received the Distinguished Science Alumni Award from Purdue in recognition of his work at MURA and later at the Synchrotron Radiation Center of the University of Wisconsin.

At MURA, Ed played an important role in the various FFAG models, particularly in the design and construction of the rf acceleration systems. After the last MURA proposal was rejected, he eagerly joined others in planning for the 240 MeV Storage Ring. In 1965, a committee of the National Academy of Science, including former MURA director P. Gerald Kruger, recommended the construction of such a ring for experimentation in the ultraviolet and near-X-ray region. When the University of Wisconsin proposal (by Mills, Snowdon, and Young) for accelerator studies was turned down by AEC, the laboratory decided to complete the ring and dedicate it to experimentation with synchrotron radiation. Ed supervised the construction of the ring and helped set up a program of experimentation at PSL. Funding for the early operation came from ARPA and OSR (Air Force Office of Scientific Research). Separate charters for PSL and the storage ring were written by the PSL Director, including a Program Advisory Committee of Users. Upon the success of these programs, the storage ring was reorganized as the Synchrotron Radiation Center (SRC), and was given the name Tantalus. This was the first dedicated storage ring for the provision of synchrotron radiation; it became the model for many such rings around the world and, importantly, introduced "big science" methods into a world of "small science" users. Applications were received from around the world for time on Tantalus. Ed showed excellent talent and judgment in carrying out the program recommended by the Program Advisory Committee and apportioning time to experimenters and experiments on Tantalus. As a result, SRC hosted a very successful and important experimental program.

Tantalus was eventually replaced by Aladdin, an 800 MeV storage ring commissioned under Ed's direction after a long and difficult struggle. On December 6, 1996, the University of Wisconsin–Madison Board of Regents approved renaming SRC the Ednor M. Rowe Synchrotron Radiation Center.

Ed and his wife, Lennie, had four children. They periodically hosted a seafood dinner with lobsters, clams, etc. shipped live on ice to Wisconsin from Maine.

10 GeV "Synchrophasotron" under construction. When completed in 1957, this mammoth, weak-focusing accelerator was the highest-energy accelerator in the world (exceeding the Berkeley 6 GeV Bevatron), and was clearly behind the administration's push for Argonne to build the ZGS. The ZGS would also be weak-focusing (the Washington administration wished to be very conservative, to the frustration of the Argonne physicists) and have an energy of 12.5 GeV, exceeding that of the Dubna machine. Dick Crane (at the University of Michigan) would occasionally refer to the ZGS as the "Stalin Memorial Accelerator" (Joseph Stalin had died in 1953). As the CERN PS (at 28 GeV) and the Brookhaven AGS (at 30 GeV) came into operation in 1959–1960, the Dubna machine's supremacy was indeed short-lived.

At Argonne, the ZGS began operation in 1963. The staff there learned that there were problems they had not anticipated. The useful aperture was not as large as had been planned, so less beam could be injected. Severe magnet current ripple modulated the extracted beam, decreasing the beam duty cycle. There were problems with the 50 MeV linac, including the preinjector and the ion source. A planned computer control system had failed to materialize, making operation awkward. Further, when the injected beam was increased the beam would disappear during the acceleration cycle. The ZGS intensity was limited to 2×10^{11} protons per pulse, with a pulse every several seconds.

MURA and its staff were asked to help solve the problems. There was some, but not universal, resistance by the ZGS staff to this idea. In any case, MURA had people who could help. Young, Curtis, and Owen helped with the linac and preinjector problems, and later helped write a proposal to upgrade the ZGS by adding a 200 MeV linac based on the MURA linac design. Peterson tried to help with magnet ripple problems, but each time he tried to perform a test, something transient would happen to cause his system to be burned up. Pruett helped diagnose the injection problems, teaching the ZGS staff how to measure tune and interpret the data. Pruett, Hilden, and Mills worked with John Martin, Russ Winje, and Tony Donaldson of the ZGS staff to address the beam loss during acceleration.

MURA had recently discovered a transverse instability (a growing amplitude of oscillation) in the beam of the 50 MeV Model that could occur at different frequencies corresponding to different modes of the beam. It had devised a feedback system that controlled all the unstable modes by using position information from each piece of the beam to deflect that piece of beam toward the beam center. Measurements by John Martin showed that the lowest-frequency modes were unstable during acceleration in the ZGS. The problem in the ZGS feedback system was that the proton velocity was changing, so the time delay between measurement and deflection changed also. Hilden devised a digital delay system that used the frequency of the acceleration system to modify the delay in order to maintain synchronism. It was also necessary to design, build, and install broadband beam position and deflection devices in the ZGS. This system, after minor adjustments, cured all the unstable modes and allowed operation at more than 10^{12} protons per pulse. A history of the ZGS by Elizabeth Paris [Paris, 2003] makes no mention of these activities. Years later, after the installation of a new vacuum tank which allowed correction of the injection magnetic field, the ZGS achieved more than 5×10^{12} protons per pulse.

After the attainment of 10^{12} protons per pulse, the University of Chicago and Argonne Lab held a great celebration for the staff, to which Mills, at least, was invited. After several hours of festivities and praise for the great job they had all done, one person, Lee Teng, in his speech said, "Oh yes, we should thank MURA also." To put all this in proper context, it is useful to recall that Albert Crewe, who was the director of the ZGS construction group and later the director of Argonne, had attacked the 1962 MURA high-intensity (2×10^{14} protons per second) FFAG proposal by claiming that the ZGS, with its two-second repetition rate, was capable of accelerating more protons per second than the FFAG on the basis of a formula "provided to him by Dr. Teng."

5.3. LINACS

The study of proton linacs of energy up to 200 MeV was initiated at MURA in 1959 at the suggestion of John Blewett, made during the summer study session dealing with the acceleration of high-intensity beams. It had been recognized even earlier that there existed no fundamental limitation on the extension of the Alvarez structure up to at least these energies. [Wilkins, 1955] The MURA studies were directed toward the use of a 200 MeV linac as an injector for the 10 GeV synchrotron. At this point in time, 50 MeV linacs with alternating gradient focusing were in early operation at BNL for the AGS, and at Harwell/CERN for the PS. Interest in extending the linac technology to Alvarez structures of greater energy intensified and in April 1961, the first Linear Accelerator Conference was held at BNL [Blewett, 1961]. By 1963, the effort became more focused at Los Alamos with a proposal to build a meson factory; at the BNL, with a proposal for a new injector for the AGS; and at MURA. It was proposed that a more collaborative detailed study be undertaken jointly by the three laboratories.

Drawing on the computational experience at MURA using mesh techniques for the calculation of magnetic fields, a program was initiated in the spring of 1960 to calculate the rf cavity fields for various cavity configurations following the techniques of J. Van Bladel and R. S. Christian [Vbladel, 1960]. A computer program called MESSYMESH was operational by late 1961 [MURA622; Young, 1963]. Refinements in the program allowed the results of a specification in the unit geometric cavity cell normalized to a frequency of 201 MHz and an accelerating field of 1 MeV/m to be included on a single page. Over 3000 geometries were computed and the results circulated to the laboratories engaged in linac design. This allowed optimization studies to be undertaken and specific linac designs to be initiated for a variety of linac applications.

Using the results of the MESSYMESH program for calculating the rf fields in the accelerating gaps of the linac, a particle dynamics computer program was developed by D. Swenson [Swenson, 1964A], [MURA714]. This program, called PARMILA, treats the phase and radial motion of the particles and allows the motion of the protons to be analyzed when they are subjected to the rf fields between drift tubes and the focusing action of the quadrupoles in the drift tubes. Such a program is necessary for the linac design to be investigated while being cognizant of beam quality and beam loss.

Linac studies require the implementation of an experimental modeling program formulated to test questionable features of the design. Three cavities resonant at a frequency of 201 MHz were constructed at MURA. The first consisted of a one-half unit cell adjustable for all energies in the linac between 1 and 200 MeV so as to allow precise frequency and field measurements to be made, mainly to determine the accuracy of the MESSYMESH field computational program. The second cavity [Young, 1966] (the sparking cavity, Fig. 5.1) was a complete unit cell

Fig. 5.1. The 200 MHz sparking cavity. It was used to determine the accelerating voltage of the structure, the critical parameter in the design of an Alvarez-type linac.

accommodating energies from 5 to 130 MeV, capable of operation at the highest power levels anticipated in the design. A value for the accelerating voltage must be chosen before the geometry of the structure can be selected to resonate at the selected frequency. A large value will shorten the length of the structure, but may result in sparking of the electrodes (drift tubes). This parameter is highly dependent on the copper surface treatment, hence the necessity of an experimental determination of a reasonable value for this parameter in consideration of the MURA experience with "clean" vacuum systems. The third cavity was a three-unit cell model at 200 MeV (Fig. 5.2), also allowing operation at full power. The fabrication of these cavities allowed mechanical engineering and fabrication problems to be studied, especially in relation to the use of copper-clad steel for the outer wall of the structure. These models proved useful later in the design, fabrication, and testing of the 200 MeV Fermilab linac injector. In fact, the three-cell cavity was later installed as a debuncher cavity in the Fermilab beam-line from the linac to the booster accelerator.

In addition to these linac studies, C. Curtis and G. Lee [Curtis, 1966] were carrying out a comprehensive program on ion source development and high-voltage (in the MeV range) accelerating-column construction. The emphasis was placed on high-intensity, low-emittance, space-charge-dominated beams.

Fig. 5.2. The three-cell, 200 MHz, 200 MeV, linac cavity that was operated at full power to confirm the mechanical, electrical, and physical design of Alvarez-type linacs at this higher energy.

After the decision was made to terminate the funding of the effort at MURA for the construction of the 12.5 GeV synchrotron, the linac studies were redirected toward a new injector for the ZGS at Argonne National Laboratory. A set of suitable parameters was selected for such a linac and more detailed studies were carried out, including engineering studies [MURA713]. When this proposal was rejected in favor of the 12 ft bubble chamber detector, the linac design effort stagnated although the participants continued to consult with the ongoing linac efforts at ANL, BNL, and LASL. A rather complete preliminary design was generated for a 100 ma, 200 MeV linac with a 10 MeV first cavity [MURA713]. This design consisted of eight accelerating cavities with a length of 128 m, excluding the drift lengths between cavities. These parameters might be contrasted with the Fermilab linac, which was constructed following the BNL specifications, with nine cavities and a length of 138 m. At the time the more conservative accelerating gradient in the BNL specifications seemed appropriate for improved reliability in the otherwise complex Fermilab accelerator complex. In fact, the extra length later proved to be an advantage, because it allowed the linac energy to be doubled from 200 MeV to 400 MeV for the same accelerator length.

In 1967, when the National Accelerator Laboratory was formed, most of the individuals, and some of the equipment, from the MURA linac group moved to the new laboratory. This allowed a jump-start on the construction of the linac injector so that the anniversary of the first 200 MeV beam in the linac was only a few days later than the 200 MeV beam in the new BNL linac.

5.4. MAGNET DEVELOPMENT

During the FFAG *époque*, Snowdon, Christian, and others had developed two-dimensional mesh relaxation magnetic field computation programs suitable for FFAG accelerators. Using these, one could rapidly try out different magnet configurations to test the properties of orbits. These were highly successful, but there were no comparable tools for conventional synchrotron magnets. Such tools were developed by Snowdon during this period. They were used to design the magnets for the Electron Storage Ring. These programs were adopted by the Lawrence Radiation Laboratory for use in the 200 BeV project. Snowdon later joined the project, by then called the National Accelerator Laboratory, and designed most of the thousands of magnets built there.

Type II superconductors were discovered in the early 1960s. They offered considerable benefit to accelerators, and in particular hadron colliders, as well as plasma containment devices, magnetohydrodynamic generators, etc. At MURA, as at many laboratories, there was intense interest in their development. This was carried out primarily by Ronald Fast and Gustavo del Castillo, in collaboration with Roger Boom, Professor of Engineering at UW. He already had considerable experience in the construction and sales of small superconducting magnets as research tools.

Several successful devices were built and tested. The design of the mountaintop cosmic ray station required large spectrometer magnets, which would need expensive power at high altitude. The principal objection to the use of superconducting magnets was the great force on the coils, and the resultant heat load of conduction through the coil support structure, which might offset the power savings from the lossless coils. A small (~30 cm) test superferric magnet was built to measure the forces and demonstrate the success of the insulation scheme.

It appeared at that time that a free-floating superconducting current hoop would be able to contribute to the stability of a plasma. In fact, the device PSL would build for Kerst would employ four free-floating current-carrying hoops (resistive or normal-conducting rather than superconducting) that were charged inductively. There were two serious problems. First, was it possible to create a permanent current which would not decay, even in the presence of a high magnetic field? Could the heat capacity of the hoop be high enough to maintain superconductivity for a useful period of time? The second problem was solved by finding a material (lead) which had a high heat capacity. The plan was to energize an outside coil to trap flux in the inner coil. The inner coil was then cooled to be superconducting, and the outside coil would be de-energized, inducing the current in the inner coil. Two versions were built, one of them with many turns of twisted multifilamentary NbTi in Cu. The induced current in this coil decayed rapidly. The second inner coil was a cylindrical multilayer with layers of Cu alternating with layers of NbSn, all layers being deposited by a plasma spray device. Using this coil it was possible to trap fields up to 3 T. Finally, the coil fractured under the stress of the trapped field.

In 1967, the experimenters left for NAL or ANL, and the AEC fusion program refused to support further work on superconducting magnets unless one could "prove that fusion was feasible with superconducting magnets," in the words of Bill Gough of AEC.

5.5. COSMIC RAYS

At the International Particle Physics Conference in Dubna in 1964, Jones, together with Fred Reines, Yash Pal, Giuseppi Cocconi, and others, were discussing their frustration with the American and European delays in moving forward with the multihundred GeV accelerators. Cocconi noted that the flux of primary cosmic ray protons above 100 GeV at mountain elevations was sufficient for exploratory studies. Back in the US, to explore this possibility, Jones got together with Fred Mills and MURA engineers Carl Radmer and Bill Winters. Fred Reines hosted a workshop at Case Western Reserve in the fall of 1964 [Case, 1964].

At MURA, a used semitrailer was equipped with a modest calorimeter, spark chambers, and related electronics. Following interaction with the physicists at the University of Denver (who administered the site), during the summer of 1965, this was taken to the summit of Mount Evans, Colorado (~4300 m elevation), where a paved road ended in a parking area and a snack/souvenir shop. This first summer was devoted to exploratory studies, understanding the cosmic ray flux and becoming familiar with the instrumentation. During the summer of 1966, the program was expanded to make a serious search for free quarks. A modest wooden structure was constructed next to the summit parking lot in which a detector array was located, which included a multilayer gas proportional chamber to measure ionization (and to be sensitive to a particle with fractional charge). The semitrailer housed the associated electronics [Jones, 1967B].

During this period, a feasibility study was carried out on a possible large, permanent mountaintop cosmic ray installation containing analyzing magnets together with large spark chambers, a liquid hydrogen target, and a calorimeter for studying in detail proton–proton interactions in the 100–1000 GeV energy range [Jones, 1966]. The MURA engineers designed the two very large magnets — one for the analysis of the incident particle (proton) above a large hydrogen target, and one below for analysis of the reaction products. Of course, these would consume an enormous 17.5 MW of power. Superconductivity was relatively new at that time; nevertheless, a superconducting model magnet was built and operated at MURA, and this option was included and discussed in the proposal. The architecture–engineering firm of Skidmore, Owings & Merrill was contracted to assist in the overall design effort. The proposed structure, over 20 m in height, would have cost US$ 23 million to construct. A proposal to build this facility on Mount Evans, signed by Jones, Mills, and Bruce Cork (then at Lawrence Radiation Laboratory), was submitted to NSF in 1967 [Jones, 1967A]. It was not funded.

The group involved in this research program was a collaboration between Jones together with Michigan technicians, students, and postdoctoral scholars, Mills with MURA (later PSL) engineers and technical staff, Don Reeder (University of Wisconsin) and students, and Bruce Cork. As the summit was inaccessible except for the few summer months, the group moved its operations to the University of Denver site near Echo Lake, accessible year round, at an elevation of 3200 m (Fig. 5.3). There the quark search experiment was continued, and a more ambitious experiment to study proton–proton interactions above 100 GeV was undertaken. For this purpose, a second, larger building was constructed, housing a detector system consisting of a 2000 L liquid hydrogen target, an iron calorimeter of about 6 m^2 area and 1130 g/cm depth, and numerous wide-gap (20 cm) spark chambers. This remains the only cosmic ray experiment to utilize a liquid hydrogen

Fig. 5.3. The building housing the Echo Lake cosmic ray experiment, at a time when liquid hydrogen was being delivered to refill the 2000 liter liquid hydrogen target. With his back toward the camera is Fred Mills.

target, which was provided by Berkeley Radiation Laboratory, thanks to Bruce Cork's involvement. It is worth noting that Bruce Cork was the nephew of James Cork, a University of Michigan faculty member who, in 1936, completed the Michigan cyclotron, the first built outside of Berkeley and, briefly, the highest-energy cyclotron in the world.

Excellent cosmic ray data were collected with this Echo Lake array during the late 1960s, including the first clean measurements of the p–p inelastic cross sections in the energy range 100–1000 GeV and other inclusive physics measurements [Jones, 1972]. The program was terminated in 1972, when the Fermilab 200 GeV accelerator was completed [Jones, 1973].

5.6. BUBBLE CHAMBERS

Cosmic rays were by no means the only involvement with particle physics experimentation undertaken by MURA. In the late 1950s, a group of Midwest physicists including William D. Walker and Myron L. (Bud) Good of the University of Wisconsin, George Tautfest of Purdue University, among others, obtained funding to build a 30" liquid hydrogen bubble chamber at MURA for general use by users at the ZGS. The MURA 30" Chamber, as it was called, was designed and constructed at MURA. It was the workhorse for the Argonne facility and performed well for many years. In 1972, it was moved to Fermilab (then National Accelerator Laboratory) and was the first operating detector there. In a few days of operation it replicated much of the data taken at Echo Lake. After the 15' bubble chamber began operation at Fermilab, the 30" chamber was decommissioned.

A second bubble chamber was proposed by Ugo Camerini and William F. Fry of the University of Wisconsin, Wilson Powell of Lawrence Radiation Laboratory, Lou Voyvodic of Argonne Laboratory, and others. This was a very large heavy liquid (freon or propane) chamber whose purpose was to find "neutral currents," i.e., interactions in which the force carrier was a heavy neutral (uncharged) boson, a heavy particle like a photon, which is a massless neutral boson. A model chamber of about 3 ft diameter and 6 ft length was constructed and operated successfully. The leadership at Argonne, which had the MURA 30" chamber, a 60" heavy liquid chamber constructed at the University of Michigan, a liquid helium chamber from Northwestern University, and was constructing a 12 ft liquid hydrogen chamber with a superconducting magnet, decided it did not need another one. CERN, however, saw the merit in the plan and constructed a chamber named "Gargamelle" which was essentially of the same design. CERN discovered neutral currents in Gargamelle — the first major discovery, among many, by that laboratory. To quote Dick Crane (Sec. 5.10), "Bad luck" — this time for Argonne.

5.7. THE ELECTRON STORAGE RING

As remarked above, in 1964, some at MURA had decided to build an electron storage ring to study alternate beam accumulation techniques, and possibly to use it as a research tool as an intense source of ultraviolet radiation. By 1965, with the ZGS activities well underway, work could begin to construct components for the ring.

The first problem was how to house it. The 50 MeV accelerator's extracted beam would be injected into the ring, so the ring had to be near to it. The accelerator was housed in an underground building set into the side of a hill, with a driveway cut into the hill leading to a large access door. The driveway had thick concrete retaining walls. Sufficient dirt cover, about 4 ft, was needed for stopping "skyshine" of the radiation (of which, because of the extremely long beam lifetime, there was very little) produced by the facility. The solution was to support precast reinforced concrete beams covering the driveway with a steel structure, which would also support a small crane for assembly purposes. The retaining walls remained, shielding was placed on top, and a concrete floor was poured, with the region under the ring isolated from the rest of the floor. The opening to the remaining driveway was sealed off with wooden framing and doors. This construction allowed the use of the same utilities and control room as those for the 50 MeV accelerator — a great saving of money. The building modification was financed by a small grant from the MURA Board.

The magnets were simple laminated C-magnet dipoles and quadrupoles. They were designed by Snowdon using the techniques he had developed for FFAG and then for synchrotron magnets. The magnet coils were fabricated by Findley Markely's group at ANL in the ZGS plastics shop. The magnets were measured and then their location chosen to minimize orbit distortions. This was perhaps the first case of sorting whole magnets to optimize accelerator performance.

The ~25 MHz rf system was an aluminum cavity coupled to the beam by a ceramic tube with metal flanges brazed thereto. It was powered by a transmitter obtained from military surplus.

The vacuum chamber was made of bakeable stainless steel, welded with a few metal gasketed flanges, pumped with ion pumps. The principal vacuum problem in such a ring, photodesorption, was not yet well understood. As a result the pressure and beam lifetime depended on intensity. Eventually, the surfaces bombarded by the photon beam cleaned up somewhat, and high-intensity operation became possible. At low intensity, the beam lifetime was many hours.

Injection was accomplished by a pulsed septum magnet and a partial aperture ceramic-loaded kicker with a fall time of several nanoseconds. Early operation was limited to single pulse injection by the lack of appropriate hardware for rf manipulation, and instability of the beam-cavity system.

It was hoped to complete the ring before the MURA Laboratory transitioned to the Physical Sciences Laboratory, but that was not possible.

5.8. THE PHYSICAL SCIENCES LABORATORY; THE SYNCHROTRON RADIATION CENTER

On July 1, 1967, the MURA Laboratory no longer existed. The University of Wisconsin Physical Sciences Laboratory (PSL) stood in its place. Bill Wallenmayer of AEC quipped to Mills at this time, "PSL — that sounds like fizzle!" It is amusing that 40 years later, PSL (and SRC) is performing vigorously. PSL was a part of the UW Graduate College on a par with the Space Sciences Center, the Army Mathematics Center, the Biotron, the Washburn Observatory, and the Computer Center. There were other laboratories as well in other colleges, such as the Engineering Research Station and the Instrumentation Systems Laboratory. Robert Bock was the Dean of the Graduate College, and therefore responsible for PSL. Bock was a very successful biologist, and he continued to perform research while Dean. It was very fortunate for MURA, and PSL, that he was in this position. If a lesser person had been in this role, things might have turned out differently.

Almost all those who would leave were gone; only one, del Castillo, moved to Argonne to work on superconducting magnets. The 200 BeV project, later called the National Accelerator Laboratory (NAL) and later still Fermi National Accelerator Laboratory, had started operation at Oak Brook, Illinois, and that part of the staff was assembling there. All the equipment needed to carry out the linac and other programs for NAL was moved to the Chicago area as soon as space to house it was found. As it developed, that included all the government-issued office furniture in the MURA Laboratory. But the University of Wisconsin had a great deal of used or unused furniture. What mattered to PSL and its staff were the tools of the Machine shop, the Electronics shop, vacuum equipment, and of course the IBM 704 computer, which still worked. In fact many of these tools were put to use on NAL projects by the PSL technical staff, of which NAL had none except those from MURA.

In the first months of operation at PSL, the lab did all it could to help the new NAL — which would soon dwarf it — get started. Within six months, NAL work tapered off as NAL found suppliers in the Chicago area (a major reason for putting it there!) and more time was spent on UW projects. Later, UW physics experimenters would use PSL to design and construct major parts of their NAL experiments. The Echo Lake experiment, started in 1965 in collaboration with L. W. Jones of the University of Michigan, continued to operate well into the 1970s. Although the large mountaintop laboratory had not been funded, the construction and operation of a more serious experiment to measure interaction cross sections of hadrons in liquid hydrogen in the energy range below 1 TeV occupied the technical, engineering, and scientific staff of PSL.

During the first year of operation of PSL, Dean Bock called upon Fred Mills, the Director of PSL, to write charters for the laboratory and its programs. PSL's mission was much broader than that of MURA. It involved many fields at UW, and was only weakly coupled to accelerators. Mills felt that the Storage Ring needed some autonomy, exactly like the other programs with which it would compete for resources at PSL. Accordingly he wrote a special charter for the Storage Ring, with the major decisions to be made by the Principal Investigator (PI) of the grant that supported it. The charter called for the formation of a users group. At the time, Mills was both Laboratory Director of PSL and PI of the Storage Ring Grant, as well as PI of several other projects at PSL. When he left UW in 1970, he recommended to OSR that they recognize Ed Rowe as the new PI of the Storage Ring, which they did. Eventually Rowe, Pruett, Otte, and Winter would name the Storage Ring "Tantalus," and their growing and successful laboratory the "Synchrotron Radiation Center" (SRC).

In the late 1970s and the 1980s, SRC would construct a 1 GeV synchrotron light source using a Microtron, designed by UW student Michael Green, as injector. UW student Walter Trzciak designed the Storage Ring lattice and undulators.

Construction of the source and facility was supervised by Rowe and Winter, while instrumentation was developed by Pruett. Pruett became an expert in ultraviolet optics and instrumentation, and built beam lines and grazing incidence focusing systems.

This separated mode of operation for the Storage Ring and the laboratory would ruffle some feathers in the early years, but both places are vibrant concerns now. They and other facilities on the former MURA site are called the "Kegonsa Campus," since the site overlooks Lake Kegonsa.

The Storage Ring had to wait to get technical support from the laboratory. Each scientist at PSL had to try to bring the benefit of PSL and their abilities to the university community. In addition, PSL took part in the University–Industry program. This was no less true for Ed Rowe, Charles Pruett, and Bill Winter than it was for Fred Mills. In early 1968, the ring had progressed to the point where injection studies could start. They arrived at a point where the beam circulated but would be kicked out by the injection kicker, which could not be shut off. Mills pried one of PSL's most skilled engineers, Frank Peterson, from his project to come up with a quick fix, which he did. When they turned on the ring again, they captured the next injected beam, accelerated it to full energy, went down to the enclosure and did minor experiments on it like moving a magnet, and watched it for several hours until they could not think of anything else to do with it. They decided to call the users and start the experimental program as soon as possible.

Rather quickly, they moved to form a users group who would perform experiments and provide advice to the laboratory about the experimental program and needed facilities. This group was chosen by nomination from the floor of the meeting and election by the group. It gave advice to the PI who was in charge of the operation and improvement of the facility. This, of course, closely parallels the management technique used at most particle physics laboratories. Thus "big science" methods worked extremely well for "small science" (whose facilities today can, of course, cost more than was spent on "big science" in the MURA era).

PSL had received a small grant from ARPA to help carry out work on the Storage Ring through that first year. Mills also applied to the Air Force Office of Scientific Research (OSR) for support, on the recommendation of ARPA. OSR was interested, but was enjoined by its creators from operating a facility for general users. On the other hand, if some UW scientists wanted to do this research, and needed the facility to do it, they could support it. And yes, it could also be used by scientists from other locations. Dean Bock helped identify several young scientists at UW who would like to receive funding and take part in the research. Two were chosen — one chemist, James Taylor, and one atomic physicist, Richard Dexter. With that, PSL received a grant which was shared with the UW users. OSR funding persisted until NSF became interested some years later. Taylor became a Distinguished Professor renowned in his field. Dexter eventually turned his knowledge of the ultraviolet to use in plasma physics, and took part in the plasma containment program. By then, the funding had shifted to NSF, and the UW researchers had their own funding.

The other major project in the first years was the Levitated Toroidal Octupole plasma containment device of Donald W. Kerst, Harold Forsen, and their students. Four current-carrying aluminum hoops of minor diameter 8″ — two weighing 935 lb, and two weighing 1650 lb — are supported briefly by a magnetic field, during which time plasma can be contained stably in the null field region. The field is caused by discharging the energy stored in a capacitor bank through series primary windings around a 50-ton laminated iron core which links the four hoops. Parallel secondary windings are interleaved with the primary windings and connected to a toroidal vacuum chamber which surrounds the hoops, and is configured to give a null field point near the centroid of the four hoops. When the field is strong enough to support the hoops, the mechanical hoop supports are withdrawn, and plasma is injected. Later, the mechanical supports recapture the hoops and return them to their original position. Students who did their thesis using this device can be found at major fusion energy laboratories, usually in leadership positions.

As time passed, PSL developed a full program [Pruett, 1972]. By 1970, the lab was self-sustaining, in the sense that all supplies and technical and scientific salaries were covered by grants, while operating costs, including the salaries of the small administrative staff, were less than the overhead received.

5.9. FERMILAB

Fermilab has been known by three names: the 200 BeV Project (1963–1967), National Accelerator Laboratory (NAL) (1967–1974), and Fermi National Accelerator Laboratory (FNAL) or simply Fermilab (1974–). The 200 BeV Project originated at Lawrence Radiation Laboratory at the University of California at Berkeley (LRL). They proposed that it be built at Camp Parks, a former military installation. Soon, however, the project was converted into a national project, for the community of particle physicists felt that the proposed management (by the University of California) of the facility would not fill the needs of the whole community. A new national corporation of universities, the Universities Research Association (URA), was formed to construct and operate the facility. A committee was appointed by the National Academy of Science to set criteria and determine the proper site for the facility. Midwest physicists, and MURA (and its successor, PSL), took part in the first and briefly the second of these periods.

While some experimenters from Midwest universities served on the main national committee, physicists and engineers served on subpanels of the committee. Fred Mills served on the Foundations and Geology Panel, chaired by Ralph

Peck of the University of Illinois, and also on the Accelerator Sciences Panel, chaired by G. Kenneth Green of Brookhaven National Laboratory. These committees rejected the Camp Parks site in California, and recommended several possible sites: Sierra Foothills, in California; Brookhaven National Laboratory, near New York City; Ann Arbor, Michigan; Stoughton, Wisconsin; and Barrington, Illinois. AEC decided to open up the site selection process to all 50 states. Ultimately about 100 sites were proposed. In order to evaluate these sites, physicists and engineers from all major laboratories and universities with accelerator experience, including MURA, were called upon by AEC to visit them. Eventually, AEC settled on the sites originally proposed, except that a site east of Batavia, including a subdivision called Weston, replaced the Barrington, Illinois site. (The citizens of Barrington decided that they were uncomfortable with strange people like physicists.)

In late 1966, AEC announced that the Weston site had been chosen. URA chose Robert R. (Bob) Wilson to be director of the new laboratory. He soon named it the "National Accelerator Laboratory." URA was to manage the project. There were several attempts to derail the project at Weston. For example, a group at Columbia University headed by Prof. Samuel Devons claimed that the result could be achieved more cheaply by locating at Brookhaven, using the AGS as injector, and dispensing with URA. That is, Associated Universities Incorporated, a corporation founded by Ivy League universities to manage Brookhaven, would manage the project. Again, MURA scientists, such as Keith Symon and Fred Mills, served on panels, or wrote letters giving advice to AEC.

In March 1967, Bob Wilson began to mobilize NAL, and called for a summer study, beginning July 1, 1967, by experts to help in the planning and design of the accelerator facility. The new PSL, which began operations on the same day, assisted as much as possible at NAL. Many MURA scientists and engineers began their transition to NAL on that day. Those who stayed at PSL took part in the summer study, did computations (NAL had no such facilities), and kept the linac program moving, while Young built a small laboratory at Downers Grove, Illinois.

In addition to supporting the linac activities, PSL built the first model main ring magnet, which was then tested and measured at Argonne. The lab also constructed a 30 m section of an elliptical stainless steel vacuum chamber, complete with ion pumps, gauges, and diagnostics. The elliptical cross section was formed from round tubing by pulling through it the same die used on the Storage Ring vacuum chamber. Excellent performance was achieved and it settled an ongoing argument at NAL with those who wanted to use diffusion pumps. PSL also did "job shopping" of small projects for NAL physicists and engineers. By early 1968, NAL had learned how to operate in the Chicago market, and needed less help, so that PSL could turn toward its future activities. Eventually, among other activities, PSL became a place where UW particle physicists built equipment for experiments at NAL (and FNAL) and elsewhere.

5.10. CYCLOTRONS AND NONSCALING FFAGS TODAY AND TOMORROW

Subsequent to the development of FFAG, spiral ridge cyclotrons were built, in a number of places, for nuclear physics studies. For example, the 88-inch cyclotron was built in the 1960s at Lawrence Radiation Laboratory (now Lawrence Berkeley National Laboratory) and this machine operated under Department of Energy sponsorship for a number of decades. In fact, at the time of this writing the machine is still operating, but is now sponsored by the Department of Energy and the national-security and space community and almost exclusively devoted to the use of ion beams to study the effect of radiation on microelectronics, optics, materials, and cells. A very large cyclotron was built at PSI (the Paul Scherrer Institut in Switzerland) with a maximum proton energy of 590 MeV ($K = 590$), a number were built at RIKEN, part of the Nishina Laboratory in Japan; and the very largest (for many decades) was at TRIUMF in Vancouver, Canada. In fact, at RIKEN they developed a set of four cyclotrons that feed one into the other. Thus rare, unstable nuclei are first produced and then accelerated. The last cyclotron in this chain, the very largest in the world, which was commissioned in December 2006, is a superconducting ring cyclotron (Fig. 5.4). At Michigan State University they built the first cyclotron using superconductivity ($K = 500$) and their $K = 1200$ machine is among the largest in the world. They have two coupled cyclotrons so as to make, and accelerate, rare ions. At the Variable Energy Cyclotron Centre in Kolkata, India, there is a superconducting $K = 520$ cyclotron. In short, there are now superconducting cyclotrons in many different places in many different countries.

Cyclotrons (in addition to other types of hadron accelerators) are also being used for cancer therapy. A superconducting cyclotron mounted on a rotating "gantry" for neutron therapy was proposed (and even patented) in the 1980s. This machine was designed, and constructed, by Henry Blosser, who had earlier interacted strongly with the MURA group. A number of the cyclotrons built for nuclear physics were also used for patient treatment. The most famous are the Harvard cyclotron and the now-decommissioned 184-inch synchrocyclotron at Berkeley. Neither of these were FFAG machines, but FFAG cyclotrons (such as the PSI machine) were, and are, used for this purpose. Over the last few decades, and increasingly in recent years, there has been considerable interest in hadron therapy for cancer, which, in some cases, can be more effective then X-ray or radioactive implant therapy. Thus we are seeing the design, construction, and use of FFAG cyclotrons built especially for cancer therapy. The firm ACCEL, an offshoot of Siemens, was bought out by Varian in the first weeks of 2007 and makes a compact

Fig. 5.4. The largest cyclotron in the world. Located at the Nishina Laboratory of the RIKEN complex in Wako, Japan, this is a superconducting, isochronous ring cyclotron. It is the largest — and final stage — of a set of four nested, cascaded cyclotrons. Here, each of the smaller, first three cyclotrons accelerates beam over an energy interval and extracts it to be injected into the next, higher energy stage.

(2.5 m diameter) machine that produces 250 MeV protons (just perfect for proton therapy). A picture of this machine is shown in Fig. 5.5. Figure 5.6 shows the Rinecker Facility in Munich, Germany, which employs the ACCEL cyclotron to supply beams to a therapy facility [Geisler, 2004].

Another use of scaling FFAGs is in driving subcritical (and hence safer) nuclear reactors. This possibility is being explored, experimentally, by Yoshiharu Mori at KURRI (a branch of Kyoto University). Two cascaded, radial sector FFAGs have been built and are shown in Fig. 5.7.

It is possible, however, to consider other types of nonscaling FFAGs which are not cyclotrons but rather are more similar to normal AG storage rings with large momentum apertures. Inevitably, this implies that the transverse betatron oscillation tune varies during the acceleration process and, hence, that resonances must be crossed. However, provided the acceleration rate is rapid enough, resonances may be crossed without excessive beam blowup. Such an FFAG was first studied for muon acceleration in either a neutrino factory or a muon collider [Mills, 1997]. Since the muon lifetime, when at rest, is only 2 μs (and still in the fractional millisecond range at 20 GeV), the acceleration must be fast. The concept has been extended to cancer therapy machines and is being studied experimentally at

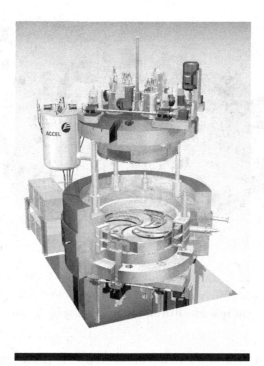

Fig. 5.5. The ACCEL K250 Superconducting Magnet Cyclotron, in use at Paul Scherrer Institute in Switzerland and the Rinecker Proton Therapy Center in Munich, Germany. (Formerly a division of Siemens Corporation, ACCEL is now owned by Varian Associates.)

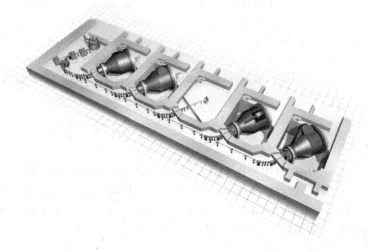

Fig. 5.6. Plan drawing of the Rinecker Proton Therapy Center, showing the cyclotron, the beam lines. There are five treatment rooms, four with gantry beam delivery and one with fixed beam direction.

Fig. 5.7. Two nested FFAG Radial Sector Machines at KURRI, Kyoto University, Japan. These machines are immediately next to a research reactor and will be used to study the performance of sub-critical reactors driven into criticality by means of accelerator beams.

Daresbury Laboratory in a device called EMMA (Electron Model for Many Applications) [Edgecock, 2007; 2008].

Nonscaling gantries (systems of beam transport magnets to deliver the beam from the accelerator to the patient, movable to point the beam in different directions), based on the FFAG nonscaling concept, have large acceptance in energy (suitable for scanning over the depth of a tumor) and very strong focusing so that the transverse size of beams is small and the variation in position with energy is also small. Thus the gantry magnets can be made very compact, and the weight of the gantry (especially if superconducting magnets are employed) is greatly reduced compared to that of the present gantries. A survey of the applications of nonscaling FFAGs may be found in the articles by Michael Craddock *et al.* [Craddock, 2004; 2008].

5.11. MURA'S LAST GASP

The last meeting of the MURA Board of Directors was on August 24, 1972. The only item on the agenda was the dissolution of the MURA Corporation. The following reminiscence is taken from a letter that Dick Crane wrote on March 2, 1994 to Frank Cole after reading the latter's *Oh Camelot!*

"I thought your last pages were an excellent analysis of why MURA lost out in the end. MURA succeeded in its goal of high current (FFAG) just at a time when the interest shifted to high energy. Bad luck.

Anyway, the main thing I want to tell you is the story of the last few hours of MURA's life, which apparently your report stopped short of. It's about the final meeting of the Board, called for the purpose of dissolving the corporation and distributing the assets.

Larry London, of U. of Minn., had been the secretary throughout the early days of MURA, and at the end he had assumed the office of chairman. He called the meeting to be held at a hotel in the O'Hare Airport area. Being our final get-together, he pulled no stops. For starters he ordered a cocktail party put on by the hotel in an upstairs room, early, I think about 4 o'clock. Larry arrived from a flight, already with a couple of sheets in the wind, from the first class service. We all got pretty happy.

A dinner table had been set up for us downstairs, where the drink service was resumed, until such a time as Larry thought we would be ready to eat. But that was not to be, for a while. Bernie Waldman had brought his wife. (I don't remember that any other wives were present.) During the proceedings, in lighting a cigarette she managed to let the whole match flap burst into flame, burning her fingers. So Bernie and one other fellow set off with her to find an emergency hospital. They were gone at least an hour, meanwhile the drink service continued at the table, and food was held off, except for snacks.

With the return of Bernie *et al.*, an elegant dinner was served. Finally with fond good-byes most of us went to airplane flights, some I believe to rooms in the hotel. A happy reunion had been had by all.

A couple of days later, in Ann Arbor, I got a call from Harold Wittig, the then-secretary of MURA, who said, "I'm calling to read you the minutes of the meeting in Chicago and get your approval." It was then the realization hit me. In the state of confusion and lubrication, we had forgotten to do what we went there to do: take formal actions. I mumbled something about how could there be minutes without a meeting. Harold said, "There are minutes now, and you can approve them or we can all go back to Chicago." I (and evidently everybody else) approved.

A result of the dissolution was about $15,000 or more to each university, from the more or less private kitty MURA had built up (evidently without the objection of AEC), largely from Marshall Keith's ingenious operations, selling time on the IBM computer and other things."

And that is the story of MURA's last gasp!

NOTE ADDED IN PROOF

David Z. Robinson, who was a member of the President's Science Advisory Committee (PSAC) at the time of this event, has written a short summary of the pertinent scientific policy issues, and his recollection about the meeting with President Johnson in the White House. Dr. Robinson has graciously allowed us to include his document in this book. We have included it as Appendix F, and express our thanks to Dr. Robinson.

— The Authors

CHAPTER 6

CONSEQUENCES AND REFLECTIONS

Before turning to the contributions, it is appropriate to reflect upon why the group was so productive. There are, clearly, a number of reasons, and the combination acted synergistically.

Firstly, MURA happened at a time in accelerator physics when strong focusing had recently been developed, and the concept of strong focusing opened a new world to accelerator physicists. This is evidenced, for example, by the independent invention of FFAG in four separate locations. Indeed, the full power of theoretical physics had not yet been applied to accelerators, and MURA was in the forefront of this new activity. It was easy for young physicists to write Hamiltonians and to use the Liouville or Vlasov equations. And it was easy for experimentalists to check these concepts and, at the same time, develop innovative approaches. The consequences for accelerator design and construction were beyond anyone's belief at that time; MURA was in the forefront of an accelerator revolution.

Secondly, computers were just becoming available for use in science, and the MURA group quickly brought the power of computers to accelerator design. That had never been done before and is, of course, the modern method of choice in accelerator design and analysis. Once again, a revolution was taking place in accelerator physics, and MURA was leading the way.

Thirdly, the group was not burdened by keeping an existing accelerator running. This freedom allowed the group to think widely, construct models, and fully engage in innovative activity.

Fourthly, an important contributor to the group's productivity was that most of the members were younger, almost all "green Ph.D.s" and almost all with no experience in accelerator design. These individuals brought fresh ideas and fresh approaches to various problems under the direction of one very senior — and very accomplished — person, Don Kerst. A great deal of the credit for the productivity of the group should thus go to him. He gently encouraged original ideas while simultaneously channeling this energy toward the concrete accomplishment of a novel but useful accelerator. To many of the young members of

MURA he was like a second thesis advisor. It was, looking back, "Jeffersonian science"* at its best.

This tradition of nurturing, teaching and training young scientists was continued throughout the existence of MURA and, certainly, was one of its most important contributions to accelerator science. In Appendix D, we list the various people associated with MURA. It is no exaggeration to say that all of those in the categories of Seniors, Staff, Engineers, and Students gained greatly from the years they spent at MURA. In the Student Sidebar we comment a bit on the various students that were involved with MURA, and then feature three of them.

6.1. INNOVATIONS

Examining the technical accomplishments, the innovations of MURA, one encounters a remarkable story. The many contributions to accelerator physics have arguably never been equaled, before or since, by any other group. It is appropriate here to both summarize the contributions and say a bit about what has happened since MURA. One sees that these contributions have revolutionized accelerator physics and that the contributions of MURA are incorporated into every modern accelerator.

Let us return to the questions raised at the beginning of this book and attempt to answer them. Of the many accelerator innovations we can think of, the most important were these.

(i) *FFAG*

Certainly, current accelerator physicists strongly associate FFAG with MURA. The MURA group pioneered the concept and built models to show that the theory was correct. The Mark V, or spiral sector, design has been employed in many cyclotrons used in nuclear physics and for medical therapy (Fig. 6.1). The general concept of FFAG is still being explored, even today (see Figs. 3.5 and 3.8), as a possible muon accelerator for a neutrino factory, as a proton driver for high-energy physics, in radioactive waste disposal, in burning thorium, and in a safe — subcritical — reactor for nuclear energy.

(ii) *Colliding beams*

A major contribution of MURA was the development of a practical method of achieving colliding beams of hadrons. The idea of colliding hadron beams had been around since before WWII, but it was the MURA group that attempted to develop the idea into something practical. Presently, essentially all new high-energy hadron machines are colliders (Figs. 6.2 and 6.3). The

---

* Jeffersonian science is a mode of research that combines aspects of the curiosity-driven and purely-applied philosophies, and investigates "… an area of basic scientific ignorance that seems to lie at the heart of a social problem." See for instance G. Holton and G. Sonnert, "A Vision of Jeffersonian Science," *Issues in Science and Technology*, Vol. 16, No. 1 (Fall, 1999), pp. 61–65.

Fig. 6.1. The TRIUMF FFAG spiral sector cyclotron at the TRIUMF Laboratory of the University of British Columbia in Vancouver, Canada. TRIUMF was for many years the world's largest cyclotron, unique in that it accelerated negative hydrogen (H^- ions) to facilitate extraction of the beam.

Fig. 6.2. A section of the Fermilab Tevatron proton–antiproton collider. Proton and antiproton beams circulate in opposite directions in the superconducting magnets near the floor against the tunnel wall.

Fig. 6.3. A section of the 27 km tunnel of the CERN Large Hadron Collider (LHC) showing the cryogenic tube containing the two counter-rotating beam lines and magnets.

concept of storage rings was developed by the MURA group, and independently by Gerry O'Neill.

(iii)   *Longitudinal manipulation of particles*

A key to the first practical hadron collider, the ISR at CERN, was the stacking of particle beams. This required understanding the effect of rf upon particles not within buckets, i.e., not engaging in simple linear oscillations. First, MURA developed an understanding of the general effect of rf upon particles, a Hamiltonian formulation of the matter, and computer numerical simulation of the effect. Thus, MURA had at its disposal the methods to understand, describe, and invent methods for the rf manipulation of particles. These methods have been the everyday tools of accelerator physicists ever since. Later hadron colliders employed these methods as well as stochastic cooling, invented at CERN, and a variety of nonadiabatic techniques invented at Fermilab and CERN, all depending on the rf theory developed at MURA. One of the methods, "slip stacking," [Mills, 1971] was used at CERN to compress the PS proton beam for antiproton production and is currently being used at Fermilab to obtain higher intensities for the neutrino program.

(iv)   *Transverse stacking*

MURA developed not only the idea of longitudinal stacking in momentum space (rf manipulation), but also the concept of transverse stacking, i.e., multiturn injection into strong focusing lattices. Once again, this contribution has become an everyday tool for accelerator physicists.

MURA Students

One of the fine things that MURA accomplished was the teaching and training of young folk. Some, who already had Ph.D. degrees when they came to MURA, although they were often still quite young, are featured in the sidebars. Other, however, actually did their Ph.D. thesis work at MURA. They are listed in Appendix D, but it seems appropriate to recognize them in a sidebar. It is not appropriate to go into detail for each of them. Below are three representative cases, of whom one went into astrophysics, one stayed in accelerator physics, and one went into fusion energy research.

David Wilkinson was a new graduate student at Michigan in 1955, when Jones and Terwilliger were constructing the Radial Sector Model. Having built an electron cyclotron as an undergraduate, he was interested in accelerators, and worked for them as a research assistant. However, in those days a doctoral thesis in accelerator science did not seem an option; also, Jones and Terwilliger were both still instructors. So Wilkinson became a student of H. R. Crane, under whom he earned his Ph.D. on the electron g-factor. He joined the faculty at Princeton University, where he devoted his career to the study of the 2.7 K cosmic microwave background radiation. A recent significant space mission in this area was named in his honor — the Wilkinson Microwave Anisotropy Probe (WMAP).

Philip L. Morton was an undergraduate and then a graduate student at Ohio State University (working under the guidance of Andrew Sessler), where he obtained a broad education in the theoretical analysis of single-particle motion in particle accelerators. His thesis was entitled "Particle Dynamics in Linear Accelerators." Subsequent to obtaining his Ph.D., he spent some time at MURA and then at Lawrence Berkeley Laboratory, before moving to SLAC, where he spent most of his career. He made many contributions to many-body phenomena in accelerators, most particularly to collective instabilities. He was one of the world's experts on the latter. His many publications range over most of the matters of interest to SLAC during his active years. They include beam positioning in Standard Positron–Electron Accelerator Rings (SPEAR), the Stanford Linear Collider (SLC) damping rings, coherent linac radiation, emittance preservation in linear colliders, and many other topics. He was even involved with the operation of the proton beam cancer therapy facility at Loma Linda.

Robert Dory attended the University of North Dakota, graduating Phi Beta Kappa in 1958, and then attended the University of Wisconsin, where he obtained a Ph.D. in 1962 under Keith Symon, specializing in Mathematical Physics and, most particularly, in particle accelerators — the business of MURA — and fusion energy. His thesis, "Nonlinear Azimuthal Space Charge Effects in Particle Accelerators," showed that above the transition energy, clumps could form, and could orbit one another in phase space. He devised a symmetry principle connecting motions above and below transition which extended his results to motions below transition, as well as a way — "Dory's rule" — to estimate the impact of the instability on the beam. Years later his work was "discovered" by others, and it was called the "Month–Messerschmidt overshoot rule." After graduation, he joined Oak Ridge National Laboratory, where he continued his work on accelerator orbit theory, and also worked on problems in plasma physics. He eventually rose to the position of head of the fusion energy theory group. He retired from Oak Ridge in 1994 and, unfortunately, suffered from a long illness leading to his death in 2006.

(v)   *Nonlinear motion*

A necessary element of FFAG, at least in the scaling version, which was the only version MURA pursued, is the use of strong nonlinear forces. Thus, MURA was "forced" into developing an understanding of nonlinear forces. Their pioneering work has become the basis for studies that have continued to this very day, with whole schools of accelerator physicists concerned with "dynamic aperture" and with designing machines to include adequately small nonlinear effects. MURA's work on Hamiltonian transformations, and its numerical work, have prompted the development of Lie transforms and other methods, which, once again, are the lifework of a whole class of accelerator physicists. MURA also discovered what came to be called "chaos" in nonlinear dynamics, and that has become a whole field of physics in itself.

(vi) *Many-body effects (static)*

Still another contribution of MURA was the careful study of static space charge effects, including the influence of beam surroundings. This theoretical work was augmented with experimental work that was, in fact, the first measurement of static space charge effects on particle beams in a cyclic accelerator. The tune shifts on both the Spiral Sector Model and the 50 MeV Model were measured with and without trapped ions.

(vii) *Many-body effects (dynamic)*

A major contribution of MURA was the first realization, and then the analysis, of a space-charge-induced beam instability. The early work on the negative mass instability became the basis for a whole field of investigation, which has continued to this day, with thousands of papers and many accelerator physicists devoting their lives to the study of wakes, electron cloud effects, head–tail instabilities, etc. It is correct to say that no modern machine would work to the intensity desired without an understanding and ability to control collective instabilities. All this grew out of the early MURA work.

(viii) *Proton linear accelerator design using digital computational techniques*

A new generation of heavy ion linear accelerators was produced using digital computational techniques familiar to the MURA group from their experience in designing magnets with complicated magnetic fields. Using digital mesh techniques that had been used in the calculation of magnetic fields, rf cavity fields were calculated for a large variety of rf cavity configurations, thus allowing geometry optimization for maximum gradient at minimum power. The technique was extended to the calculation of beam dynamics. These digital computational techniques were shared and copied by other laboratories engaged in the design of the new generation of heavy ion linear accelerators.

(ix) *Storage rings for synchrotron radiation research*

The MURA group went on to develop the use of the electron storage ring for the production of synchrotron radiation for research. Then PSL completed

Henry G. Blosser (1928–)

Henry G. Blosser was educated at the University of Virginia, obtaining a B.A. in 1951, an M.S. in 1952, and a Ph.D. in 1954. Immediately after that he went to Oak Ridge National Laboratory and worked under Ted Welton. Only two years later he was made Group Leader of Cyclotron Analogue One. It was during that time that he became involved with MURA (even spending some time there) and, especially, with the concept of FFAG as applied to cyclotrons. In fact, most of his professional life since then has been devoted to continuous wave FFAG cyclotrons.

In 1958, Blosser moved to Michigan State University (MSU), and quickly worked through the ranks, becoming a full professor in 1961 and, finally, University Distinguished Professor in 1990. He was instrumental in making MSU the center for cyclotrons and nuclear physics, having been involved with the design and construction of all four MSU cyclotrons. While being a professor he was Director of the Cyclotron Laboratory from 1958 to 1966 and, again, from 1969 to 1985. Presently he is Adjunct Professor in the Department of Radiation Oncology at Wayne State University.

His areas of expertise include the design/development of advanced particle accelerator systems, superconducting magnets for cyclotrons, apparatus for nuclear physics research, and accelerators for cancer therapy. He is the holder of 5 patents and has published more than 70 scientific papers. He has been asked to speak more than 40 times, and has presented more than 50 scientific papers, at national/international conferences.

Blosser has, of course, been the person most responsible for the excellent nuclear physics (which is still going on) at MSU. He also pioneered the use of superconductivity in cyclotrons. At the same time he pioneered the development of cyclotrons for cancer therapy. It is interesting how he got into this important activity. In his own words:

"The creation of the superconducting medical cyclotron in Detroit came about due to a reporter calling me and asking what superconducting cyclotrons were good for. One of the things I said in responding was that they were good for cancer treatments. Dr. William Powers, M.D., happened to be listening to the program and called me and said, 'I need one of those cyclotrons.' In a matter of a few months money appeared, supplied by two widows (Bernice and Sylvia Gershenson) who had lost their husbands due to cancer. So we built a superconducting cyclotron for Harper Hospital that was small enough and light enough to be mounted on a gantry and which could rotate thru a full 360° about a patient on a horizontal table. By 1991, a steady stream of patients was being treated quite successfully."

Since that time Blosser has been involved with the design of a number of cyclotrons for cancer therapy. Most particularly, Blosser and MSU, in the year 2000, entered into a formal

(Continued)

Henry G. Blosser (*Continued*)

agreement with the German company ACCEL Instruments to assist in the design and construction of superconducting medical accelerators producing 250 MeV protons, based upon work Blosser had done in 1993. ACCEL was formed from Interatom, the nuclear branch of Siemens Corporation, which proposed both a synchrotron and a superconducting cyclotron version for the Massachusetts General Hospital's proton therapy facility. ACCEL is now a part of Varian Instruments Inc.

Besides his professional career — he still goes into the Laboratory essentially every day Blosser has raised a family, likes hiking and biking, and travels a good deal.

its construction and operated it successfully under Ednor Rowe and Charles Pruett. The highly successful Synchrotron Radiation Center at the University of Wisconsin is an outgrowth of this work. The very modest investment in the storage ring and the introduction of "big science" techniques into "small science" have led the way to a billion-dollar-plus research field with dozens of facilities worldwide.

Some other contributions could be mentioned, and have been mentioned in the body of this report, but perhaps enough has been presented here to convince the reader of the many technical contributions of MURA. And yet, despite the prolific achievements and the technical competence of the group, no large accelerator ever came to MURA.

6.2. INNOVATION WAS NOT ENOUGH

The primary goal of MURA's formation had been to bring a high-energy accelerator and its laboratory to the Midwest under the aegis of the combined Midwestern universities. In order to understand why MURA failed in this, we must place its activities in the proper context. In 1945, at the end of World War II, the US government had in place the Manhattan District, a military project that had developed the atomic bomb. Some of the major assets were the reactors at Hanford, Washington, which produced the plutonium for weapons; the two plants at Oak Ridge, Tennessee to separate uranium 235 from natural uranium by thermal diffusion and electromagnetic means; the Los Alamos Laboratory, which had designed and tested the weapon; the Radiation Laboratory at the University of California at Berkeley; the Reactor Group at the University of Chicago (later called Argonne

Fishing Trips

The MURA Laboratory was located in Madison, Wisconsin, a place of scenic beauty in the midst of many opportunities for recreation. MURA staff members took advantage of these opportunities — for example, on fishing trips organized by Marshall Keith at Lake Riley in the Nicolet National Forest, near his hometown of Crandon, Wisconsin. Lake Riley is accessible only by boat from Lake Wabikon, poling the motorboats upstream through the connecting stream. Because of its remoteness, fishing was quite good on occasion, yielding perch, bass, northern pike, and the rare muskellunge, as in the first figure. Tents were pitched, as in the second figure, on a promontory by the lake, and meals were cooked over an open fire, as in the third figure. The trips were usually taken in the early autumn, when the mosquito and black fly populations were reduced.

On some trips the weather was quite bad, with strong winds, rain, and lightning. During one such storm a muffled cry was heard, "Help! Help!" "What's wrong?" "I'm a prisoner of Don Swenson's tent! Help!" It was Phil Meads, who had borrowed a tent from Swenson, and the tent had blown down, and he could not get out. The next morning, as we shivered around the fire waiting for the coffee to boil, we heard a strange animal sound like "Urp! Urp!". Investigation showed that it was Meads swimming naked in the lake, doing a good job of imitating a sea lion. On another trip, taken in October, the temperature dropped to 19° F. Waves splashed water into the boats, which then made icicles. After not very much fishing, and a lot of time in the tents, the group left, motoring through 1/4 in. of ice on Lake Wabikon.

On one of the trips, Frank Peterson and Don Swenson went early and arrived before anyone else. According to their story, they encountered a deer swimming in Lake Riley and decided to lasso it with the anchor rope for some purpose. While it is one thing to lasso a swimming deer, it is quite another thing to get the rope back without swamping the boat. Eventually they succeeded, but not without difficulty. When others arrived, there was a tendency to doubt the story's verity. Nevertheless, almost 50 years later, they insist that the story was true.

Don Kerst returned to the University of Wisconsin–Madison (UWM) in 1962. Although he did not return to MURA, he kept in touch. Symon, Kerst, Mills, and Raymond Herb

(Continued)

Fishing Trips (*Continued*)

of UWM and families took a series of canoe trips on Wisconsin rivers, such as the Flambeau River in the Flambeau River State Park, the Saint Croix Scenic Riverway, the Brule River in the Brule River State Forest, and the Wisconsin River near Spring Green. All trips were camping trips, and the first three required shooting some serious rapids, while the last was a simple float trip, with camping on sandbars in the middle of the river. Later, others, Don Young in particular, started canoe trips more in the style of the fishing trips. These trips continued in Don's linac group at Fermilab for some years after leaving MURA. In August 1967, Snowdon and Mills organized the "First NAL Canoe Trip" on the Flambeau River. Among those attending were Ernest Courant of BNL and Alpert Garren of LBL, as well as some employees of DUSAF, the architecture–engineering firm building NAL. It rained the first night, and it was a memorable thing to see Ernest using an air mattress pump to coax the fire into life to make coffee while Joyanne Mills prepared to cook breakfast.

Besides fishing and canoeing trips, there were, of course, many other social activities of the MURA group. As soon as they arrived in Madison, the Purdue group — Haxby, Wallenmayer, and Rowe — set up a coffee group, apparently following their practice at Purdue. Hot water, instant coffee, tea bags, sugar, and creamer were offered for a nominal fee (probably a dime). Rather quickly, a substantial amount of money was stored up and was used to fund a picnic, to which all were invited, with their spouse or friend. Bratwurst and beer were the main attractions, but more "family-friendly" food was also served. The "Coffee Fund Picnics" were continued at the Physical Sciences Laboratory after 1967. MURA was a place where people liked those they worked with, and welcomed the chance to socialize or recreate with their colleagues. There are many other examples, such as waterskiing in the summer, ski trips at −20°F at Don and Bille Young's cabin, or ladies' teas initiated by Mona Haxby.

Laboratory); the Rare Earth Laboratory at Iowa State University; and many other fabrication and engineering laboratories.

In large part due to the efforts of David Lilienthal, together with the senior scientists of the project, the management of this project and these facilities was transferred out of the War Department and into civilian hands. The Atomic Energy Commission (AEC) was created, and began operation January 1, 1947. There was a perceived problem with transitional funding in the university laboratories. The Radiation Laboratory at Berkeley had been in operation for some years before WWII, and continued as an internal laboratory of the University of California. Argonne Laboratory had no such history. Funding the laboratory in the transition period was beyond the ability of the University of Chicago. On the other hand, reactors were the principal method for the commercial exploitation of atomic energy, so that Argonne Laboratory was highly valued by the Midwestern universities. Accordingly a group of these universities agreed to underwrite the salaries of the

Argonne employees in case timely financing from the government was not available. As it happened, the federal funding was received in time.

The University of Chicago negotiated a contract with AEC similar to that at Berkeley. That is, the laboratory was an arm of the University of Chicago and was not available for research by scientists from other institutions. As late as 1957, Mills heard Jack Livingood (head of the Accelerator Project at Argonne) tell a Purdue University professor (George Tautfest) that if he wanted to do research with the ZGS he had better fill out a job application for Argonne. The interests of Midwestern nuclear physicists turned away from reactors to particle accelerators. This led, eventually, to the Midwest Accelerator Conference and MURA. In 1955, upon learning that the Soviet Union was undertaking the construction of a 10 GeV weak focusing proton synchrotron, AEC, in an act of inspired (but misguided) leadership, asked its reactor laboratory to build a 12.5 GeV weak focusing proton synchrotron, called the Zero Gradient Synchrotron (ZGS), at great speed in time to "beat the Russians." Apparently AEC did not have confidence that the Brookhaven Alternating Gradient Synchrotron, already under construction, would ever work satisfactorily. (In fact it worked as expected and came in before the ZGS and with higher energy and even higher intensity.) When this was announced, AEC, in particular Willard Libby, encouraged the MURA Laboratory to continue the development of advanced accelerator concepts.

With the advent of the ZGS, the only way MURA could do this was by creating something exceptional. The only real possibilities were higher energies or higher intensities. It first explored the idea of higher energies by colliding beams. The use of the Two-Way FFAG, although providing higher energies, did not lead to a clear way to carry out experiments. Further, it would develop that synchrotrons could reach similar energies (500 GeV) and perform experiments in ways already explored. When fitted with all the embellishments developed at MURA, the synchrotron could reach a similar rate of particle production to the lower-energy FFAG. Similarly, the center-of-mass energies reachable with colliding beams in FFAGs could be reached by higher-energy synchrotrons, and they offered a better way to study the physics. (This point was made at the 1959 MURA Study.) The J/psi particle was discovered both at BNL in a primary proton beam and at SLAC in SPEAR, an e^+–e^- collider; ADONE, an e^+–e^- collider with 1.5 GeV in each beam, ran for many years only 50 MeV below the energy that would have produced the J/psi, and they never saw it.

Thus, MURA found itself with the wrong tools to tackle the day's particle physics problems. As an example of this, note that the B quark was found at the Fermilab Main Ring, but not at the ISR, which had higher center-of-mass energy. Later the point would be reached where the desired energy could only be achieved by colliding beams. Thus, no one has ever proposed a proton synchrotron of sufficient energy to be a "top" factory, or even a Z factory — both those particles were discovered at colliders — except possibly Enrico Fermi, in his famous 1955 talk at the American Physical Society, in which he described an accelerator circling the earth.

Looking back, it appears that MURA did not have the external advice that might have made its proposals more compelling. At least, review committees provide communication channels to other laboratories and active physicists as well as a recognized body of experts whose opinions are taken seriously. Perhaps such a system of committees would have made a big difference. No one will ever know.

It has been stated that the demise of MURA promoted the construction of the 200 GeV accelerator in the Midwest as its due. It is also possible that politicians noticed the large price tag for that accelerator, so that politics followed its normal course in the selection of the site and the former existence of MURA had little to do with the site selection process.

6.3. PERSONAL NOTE

In conclusion, it is fitting to remember the likening of MURA to Camelot by Frank Cole and Mervin Hine (Fig. 6.4). Those of us fortunate enough to have been a part of MURA do remember it as being exactly that way.

Fig. 6.4. The MURA alumni at the 1989 Symposium honoring the memory of Kent Terwilliger. Standing, L-R: Donald Swenson, Ednor Rowe, Donald Young, Keith Symon, H. Richard Crane, Charles Pruett, and Frank Cole. Seated, L-R: Fred Mills, Andrew Sessler, Lawrence Jones, Ernest Courant, Donald Kerst, and William Wallenmeyer.

BIBLIOGRAPHY

Bibliography items are identified by a label consisting of an author's name, followed by the year of publication, possibly followed by a capital letter if the author has more than one item in a given year. They are listed below alphabetically by this identity label. An item is referred to in the text by its identity label enclosed in square brackets.

Many of the conference proceedings can be found at the Joint Accelerator Conferences Website (JACoW) http://accelconf.web.cern.ch/accelconf

Other conference proceedings can be found in the Clearinghouse for Federal Scientific and Technical Publications, National Bureau of Standards, US Department of Standards.

Arnold 1978	V. I. Arnold, *Mathematical Methods of Classical Mechanics* (Springer, New York, 1978).
Barber 1966	W. C. Barber *et al.*, Operation of the electron–electron storage ring at 550 MeV, *Symp. Int. Sur Les Anneaux De Collisions* (Orsay, Saclay, 1966) ed. H. Zynger and E. Cremieu-Alcan, Vol. II, p. 2.
Barton 1961	M. Q. Barton and C. E. Nielsen, Longitudinal instability and cluster formation in the Cosmotron, *Proc. 1961 Int. Conf. High Energy Accelerators*, (Brookhaven National Laboratory, 1961), p. 163.
Blewett 1956	M. H. Blewett, CERN Symposium on High Energy Physics, *Phys. Today* **9**(11), 18 (1956).
Blewett 1961	J. P. Blewett (chairman), Minutes of the Linear Accelerator Conference (BNL, April, 1961), AvS-1.
Budker 1956	G. J. Budker, Relativist stabilized electron beam, *CERN Symp. High Energy Accelerators and Pion Physics* (1956), p. 68.
Budker 1966	G. I. Budker *et al.*, Electron–electron scattering at 2x135 MeV, *Symp. Int. Sur Les Anneaux De Collisions* (Orsay, Saclay, 1966) eds. H. Zynger and E. Cremieu-Alcan, Vol. 2.

Case 1964   L. Jones, A. Pevsner and F. Reines, *Proc. Conf. Interaction Between Cosmic Rays and High Energy Physics* (Case Institute of Technology, 1964).

Chirikov 1979   B. V. Chirikov, *Phys. Rep.* **52**(5), 263 (1979).

Christian 1961   R. S. Christian, C. D. Curtis, G. M. Lee, F. E. Mills, F. L. Peterson, C. A. Radmer, M. F. Shea and D. A. Swenson. Injection into the MURA 50 MeV electron accelerator, *Bull. Am. Phys. Soc.* **6**, 447 (1961).

Christofilos 1950   N. C. Christofilos, US Patent 2,736,799 (1950).

Cole 1955   F. T. Cole and D. W. Kerst, Small model FFAG betatron II, design calculations, *Phys. Rev.* **100**, 1246 (1955).

Cole 1957   F. T. Cole *et al.*, Electron model fixed field alternation gradient accelerator, *Rev. Sci. Instrum.* **28**, 403 (1957).

Cole 1959A   F. T. Cole and P. L. Morton, Radial straight sections in spiral sector FFAG accelerators, *CERN Symp. Proc.* (1959), pp. 31–37.

Cole 1959B   F. T. Cole, Typical designs of high energy FFAG accelerators, *CERN Symp. Proc.* (1959), pp. 82–88.

Cole 1994   F. T. Cole, "Oh Camelot!: A Memoir of the MURA Years", (April, 1, 1994), unpublished.

Congress 1965   High Energy Physics Program: Report on National Policy and Background Information of the Joint Committee on Atomic Energy, Congress of the United States, February 1965.

Courant 1952   E. D. Courant, M. S. Livingston and H. S. Snyder, The strong-focusing synchrotron — A new high energy accelerator, *Phys. Rev.* **88**, 1190 (1952).

Craddock 2004   M. Craddock, The rebirth of the FFAG, *CERN Courier* **44** (6), Art. 17 of 21, July/August, 2004.

Craddock 2008   M. K. Craddock and K. R. Symon, Cyclotrons and Fixed-Field Alternating-Gradient Accelerators, *Reviews of Accelerator Science and Technology* Vol. 1, ed. A. W. Chao and W. Chou (World Scientific, 2008).

Curtis 1963   C. D. Curtis, A. Galonsky, R. H. Hilden, F. E. Mills, R. A. Otte, G. Parzen, C. H. Pruett, E. M. Rowe, M. F. Shea, D. A. Swenson, W. A. Wallenmayer and D. E. Young, Beam experiments with the Mura 50 MeV FFAG accelerator, *Proc. Int. Conf. High Energy Accelerators* (Dubna, USSR, 1963), p. 620.

Curtis 1964   C. D. Curtis and G. E. Lee, Preaccelerator column
                     design, *Proc. Linear Accelerator Conference at MURA* (1964),
                     pp. 487–496.

Curtis 1966   C. D. Curtis, G. M. Lee and J. A. Fasolo, Initial tests of
                     the MURA high gradient column, *Proc. 1966 Linear
                     Accelerator Conference*, Los Alamos Report LA-3609
                     (1966), pp. 365–370.

Edgecock 2007   R. Edgecock (EMMA Collaboration), EMMA: The
                     World's First non-scaling FFAG, *Proc. Particle Accelerator
                     Conference* (Albuquerque, New Mexico, 2007).

Edgecock 2008   R. Edgecock *et al.*, EMMA: The world's first non-scaling
                     FFAG, *Proc. European Accelerator Conference* (2008).

Ferger 1963   F. A. Ferger *et al.*, The CERN electron storage ring
                     model, *Int. Conf. High Energy Accelerators* (Dubna, 1963).

Geisler 2004   A. Geisler *et al.*, Status report of the ACCEL 250 MeV
                     medical cyclotron, *Proc. Cyclotrons and Their Applications*
                     (Tokyo, 2004) ed. A. Goto, p. 178.

Greenberg 1999   D. S. Greenberg, *The Politics of Pure Science* (University of
                     Chicago Press, 1999).

Greenbaum 1971   L. Greenbaum, *A Special Interest: The Atomic Energy
                     Commission, Argonne National Laboratory and the Midwestern
                     Universities* (University of Michigan Press, Ann Arbor, 1971).

Guignard 1989   G. Guignard, Non-linear dynamics, *AIP Conf. Proc.* **184**,
                     820 (1989).

Haxby 1959   R. O. Haxby, L. J. Laslett, F. E. Mills, F. L. Peterson,
                     E. M. Rowe and W. A. Wallenmeyer, Experience with a
                     spiral sector FFAG accelerator, *CERN Symp. Proc.* (1959),
                     pp. 75–81.

Haxby 1961   R. Haxby *et al.*, Spiral-sector FFAG magnets, *BNL
                     Accelerator Conf.* (1961), pp. 476–478.

Hoddeson 2008   L. Hoddeson, A. W. Kolb and C. Westfall, *Fermilab*
                     (University of Chicago Press, 2008).

Johnsen 1971   K. Johnsen, The CERN intersecting storage rings, *Proc.
                     8th Int. Conf. High Energy Accelerators* (CERN, 1971), p. 79.

Johnstone 1997   C. Johnstone, FFAG non-scaling lattice designs, *Proc.
                     4th Int. Conf. Physics Potential of $\mu^+ - \mu_-$ Colliders* (San
                     Francisco, December, 1997), pp. 696–698.

Jones 1954   L. W. Jones, "Patch" operation of a fixed field alternating
                     gradient accelerator, Mark III, with comments on

momentum compaction and amplitude of oscillation, *MURA* **51** (1954).

Jones 1955A — L. W. Jones, K. R. Symon, K. M. Terwilliger and D. W. Kerst, Synchrotron application of reversed field types of fixed field alternating gradient magnets, *Phys. Rev.* **98**, 1153 (1955).

Jones 1955B — L. W. Jones and K. M. Terwilliger, Small model FFAG betatron III: Preliminary experimental results, *Phys. Rev.* **100**, 1247 (1955).

Jones 1956A — L. W. Jones and K. M. Terwilliger, A small model fixed field alternating gradient radial sector accelerator, *CERN Symp. High Energy Accelerators and Pion Physics* (CERN, Geneva, 1956), Vol. 1, pp. 359–365.

Jones 1956B — L. W. Jones, K. M. Terwilliger and R. O. Haxby, Experimental test of the fixed field alternating gradient principle of particle accelerator design, *Rev. Sci. Instrum* **27**, 651 (1956).

Jones 1957A — L. W. Jones, K. M. Terwilliger and C. H. Pruett, Betatron oscillation resonances in a small FFAG betatron, *Bull. Am. Phys. Soc.* **11**(2), 10 (1957).

Jones 1957B — L. W. Jones and K. M. Terwilliger, Radio-frequency with a FFAG accelerator I: Methods, *Bull. Am. Phys. Soc.* **11**(2), 188 (1957).

Jones 1957C — L. W. Jones, Notes on the Berkeley Budker Conference. MURA Report No. 364 (1957).

Jones 1959A — L. W. Jones, C. H. Pruett, K. R. Symon and K. M. Terwilliger, Comparison of experimental results with the theory of radio-frequency acceleration processes in FFAG accelerators, *Int. Conf. High-Energy Accelerators and Instrumentation* (CERN, Geneva, 1959), pp. 58–70.

Jones 1959B — L. W. Jones, Experimental utilization of colliding beams, *CERN Symp. Proc.* (1959), pp. 15–22.

Jones 1959C — L. W. Jones, C. H. Pruett, K. R. Symon and K. M. Terwilliger, Comparison of experimental results with the theory of radio-frequency acceleration processes in FFAG accelerators, *CERN Symp. Proc.* (1959), pp. 58–70.

Jones 1959D — L. W. Jones and K. M. Terwilliger, Beam extraction from FFAG synchrotrons, *CERN Symp. Proc.* (1959), pp. 48–57.

Jones 1963A — L. W. Jones, Recent US work on colliding beams, *Proc. Int. Conf. High Energy Accelerators* (Dubna, 1963), pp. 300–311.

Jones 1963B	L. W. Jones and B. de Raad, Experimental utilization of proton storage rings, *Nucl. Instrum. Meth.* **20**, 477 (1963).
Jones 1965	L. W. Jones, The use of cosmic rays to study physics in the range 100–1000 GeV, *Proc. Int. Conf. High Energy Accelerators* (Frascati, 1965).
Jones 1966	L. W. Jones, The use of cosmic rays to study physics in the range 100–1000 GeV, *V Int. Conf. High Energy Accelerators* (Frascati, 1965; CNEN, Rome, 1966), p. 66.
Jones 1967A	L. W. Jones, F. E. Mills and B. Cork, Proposal to the National Science Foundation for untra high energy cosmic ray physics facility, University of Michigan, ORA, 67-1588-F1 (1967).
Jones 1967B	L. W. Jones *et al.*, Search for massive particles in cosmic rays. *Phys. Rev.* **194**, 1584 (1967).
Jones 1972	L. W. Jones *et al.*, The properties of proton–proton interactions between 100 and 1000 GeV from a cosmic ray experiment, *Nucl. Phys.* **B43**, 477 (1972).
Jones 1973	L. W. Jones, The history, highlights and outcome of the Michigan–Wisconsin Echo Lake Cosmic Ray Program, 1965–1972: An informal review, Unpublished University of Michigan Technical Report, UM HE 73-9 (1973).
Jones 1991	L. W. Jones, Kent M. Terwilliger; graduate school at Berkeley and early years at Michigan, 1949–1959, *Kent M. Terwilliger Memorial Symposium. AIP Conf. Proc.* **237**, 1 (1991).
Kaiser 2007	D. Kaiser, Viki Weisskopf: Searching for simplicity in a complicated world, *MIT Phy. Ann.* (2007), p. 44.
Kapchinskij 1959	M. Kapchinskij and V. V. Vladimirskij, Limitations of proton beam current in a strong focusing linear accelerator associated with the beam space charge, *Int. Conf. High-Energy Accelerators and Instrumentation* (CERN, 1959), p. 274.
Kerst 1948	D. W. Kerst, A process aiding the capture of electrons injected into a betatron, *Phys. Rev.* **74**, 503 (1948).
Kerst 1955A	D. W. Kerst, Contant frequency cyclotrons with spirally ridged poles, MURA Report No. 64, March, 29 (1955).
Kerst 1955B	D. W. Kerst, K. M. Terwilliger, K. R. Symon and L. W. Jones, Fixed field alternating gradient accelerator with spirally ridged poles, *Phys. Rev.* **98**, 1153 (1955) (abstract).
Kerst 1956A	D. W. Kerst, K. R. Symon, L. J. Laslett, L. W. Jones and K. M. Terwilliger, Fixed field alternating particle accelerators, *CERN Symp. Proc.* **I**, 366 (1956).

Kerst 1956B   D. W. Kerst, Spiral sector magnets, *CERN Symp. Proc.*
                   I, 32 (1956).
Kerst 1956C   D. W. Kerst, F. T. Cole, H. R. Crane, L. W. Jones,
                   L. J. Laslett, T. Ohkawa, A. M. Sessler, K. R. Symon,
                   K. M. Terwilliger and N. Vogt Nilsen, Attainment of very
                   high energy by means of intersecing beams of particles,
                   Phys. Rev. **102**, 590 (1956).
Kerst 1956D   D. W. Kerst, Properties of an intersecting-beam
                   accelerating system, *CERN Symp. Proc.* **1**, 36 (1956).
Kerst 1957   D. W. Kerst and F. E. Mills, Injection in the spiral
                   sector FFAG Model accelerator, *Bull. Am. Phys. Soc.* **2**,
                   337 (1957).
Kerst 1960   D. W. Kerst, E. A. Day, H. J. Hausman, R. O. Haxby,
                   L. J. Laslett, F. E. Mills, T. Ohkawa, F. L. Peterson,
                   E. M. Rowe, A. M. Sessler, J. N. Snyder and
                   W. A. Wallenmayer, Electron model of a spiral sector
                   accelerator, *Rev. Sci Instrum.* **31**(10), 1076 (1960).
Kerst 1985   Fermilab conference paper (1985).
Kerst 1989   D. W. Kerst, Accelerators and the Midwestern
                   Universities Research Association in the1950s, *Pions to
                   Quarks: Particle Physics in the 1950s* (based on a 1985
                   Fermilab symposium), eds. L. M. Brown, M. Dresden,
                   L. Hoddeson and M. West (Cambridge University Press,
                   1989), pp. 202–212.
Kerst 1991   D. W. Kerst, Terwilliger and the group: A chronicle of
                   MURA, *Kent M. Terwilliger Memorial Symposium, AIP
                   Conf. Proc.*, eds. L. W. Jones and A. D. Krisch (American
                   Institute of Physics, New York, 1991), Vol, 237, pp. 22–52.
Kinraide 2000   R. Kinraide, Report on the MURA Project, Physical
                   Sciences Laboratory, University of Wisconsin–Madison,
                   January 13, 23000.
Kolomenskij 1958   A. A. Kolomenskij, A symmetric circular phasotron with
                   oppositely directed beams, *Sov. Phys. JETP* **6**, 231 (1958);
                   English version of 1957 paper in Russian.
Kolomenskij 1959   A. A. Kolomenskij and A. N. Lebedev, Certain beam-
                   stacking effects in fixed-field magnetic systems, *Proc. Int.
                   Conf. High-Energy Accelerators and Instrumentation* (CERN,
                   1959), ed. L. Kowarski.
Kriegler 1964   F. J. Kriegler, Dynamic effects of higher RF harmonics, *Proc.
                   Linear Accelerator Conference at MURA* (1964), pp. 204–213.

Laslett 1956 L. J. Laslett and K. R. Symon, Particle orbits in fixed field alternating gradient accelerators, *CERN Symp. Proc.* **1**, 279 (1956).

Laslett 1959A L. J. Laslett and K. R. Symon, Computational results pertaining to use of time-dependent magnetic field perturbation to implement injection or extraction in a FFAG synchrotron, *CERN Symp. Proc.* (1959), pp. 38–47.

Laslett 1959B L. J. Laslett and K. R. Symon, Computational results pertaining to use of a time-dependent magnetic field perturbation to implement injection or extraction in a FFAG synchrotron, *Proc. Int. Conf. High-Energy Accelerators and Instrumentation* (1959), pp. 38–47.

Laslett 1961 L. J. Laslett, V. K. Neil and A. M. Sessler, Coherent electromagnetic effects in high-current particle accelerators III: Electromagnetic coupling instabilities in a coasting beam, *Rev. Sci. Instrum.* **32**(3), 276 (1961).

Laslett 1963A L. J. Laslett, Particle motion in the proposed Budker accelerator, *Dubna Conf. Proc.* (1963), pp. 1042–1048.

Laslett 1963B L. J. Laslett, *Proc. 1963 Summer Study on Storage Rings*, BNL Report 7534 (1963), p. 324.

Laslett 1965 L. J. Laslett, V. K. Neil and A. M. Sessler, Transverse resistive instabilities of intense coasting beams in particle accelerators, *Rev. Sci. Instrum.* **36**, 436 (1965).

Lichtenberg 1956 D. B. Lichtenberg, R. G. Newton and H. M. Ross, Intersecting beam accelerator with storage ring, *MURA Rep.* **110** (1956).

Livington 1962 M. S. Livingston and J. P. Blewett, *Particle Accelerators* (McGraw-Hill, New York, 1962).

Martin 1965 J. H. Martin, R. A. Winje, R. H. Hilden and F. E. Mills, Damping of the coherent vertical beam instability in the Argonne ZGS, *Proc. Fifth Int. Conf. Particle Accelerators* (Frascati, Italy; September 9–16, 1965).

Maxwell 1856 J. C. Maxwell, *Scientific Papers* (Cambridge University Press, 1890), Vol. I, p. 288 (Adams Prize Essay, 1856).

Meier 1959 H. Meier and K. R. Symon, Analytical and computational studies on the interaction of a sum and a difference resonance, *CERN Symp. Proc.* (1959), pp. 253–261.

Michelotti 1989 L. Michelotti, Phase space concepts, *AIP Conf. Proc.* **184**, 891 (1989).

Mills 1959	F. E. Mills and T. O. Binford, Injection into accelerators by a programmed perturbation, *Bull. Am. Phys. Soc.* **4**, 268 (1959).
Mills 1961A	F. E. Mills and D. C. Morin, Multiturn injection into FFAG accelerators, *Proc. Int. Conf. High Energy Accelerators* (Brookhaven, USAEC, 1961), pp. 395–400.
Mills 1961B	F. E. Mills, J. A. Mogford, C. A. Radmer and M. F. Shea, Beam extraction from an FFAG accelerator, *BNL Accelerator Conf.* (1961), pp. 417–421.
Mills 1963A	F. E. Mills *et al.*, Beam experiments with the MURA 50 MeV FFAG accelerator, *Dubna Conf. Proc.* (1963), pp. 620–651.
Mills 1963B	F. E. Mills *et al.*, Beam extraction from the MURA 50 MeV FFAG accelerator, *Dubna Conf. Proc.* (1963), pp. 723–729.
Mills 1963C	F. E. Mills, Classical radiation damping in accelerators, *Nucl. Instrum. Meth.* **23** 197 (1963).
Mills 1965A	F. E. Mills, R. A. Otte and C. H. Pruett, Diagnostics and control of a coherent instability in the MURA 50 MeV electron accelerator, *Proc. Int. Conf. High Energy Accelerators* (Frascati, 1965), pp. 343–347.
Mills 1965B	F. E. Mills, R. H. Hilden, J. H. Martin and R. J. Winje, Damping of the coherent vertical instability in the ZGS, *Proc. Int. Conf. High Energy Accelerators* (Frascati, 1965), pp. 347–350.
Mills 1965C	F. E. Mills, Coherent acceleration of protons, *Proc. Int. Conf. High Energy Accelerators* (Frascati, 1965), pp. 444–447.
Mills 1971	F. E. Mills, Stability of phase oscillations under two applied frequencies, AADD-176 Brookhaven National Laboratory Internal Report (1971).
Mills 1997	F. E. Mills, Linear orbit recirculators, *Proc. 4th Int. Conf. Physics Potential of $\mu^+ - \mu^-$ Colliders* (San Francisco, December 1997), pp. 693–695.
Mills 2001	F. E. Mills, *CYCLOTRONS2001 Conference*.
Moser 1955	J. Moser, Stabilitätsverhalten kanonischer differentialgleichungssystems, *Nach. Akad. Wiss. Göttingen. Math. Phys. Kl.* (1955), pp. 87–120.
MURA 1962	MURA staff, Design of a 10-BeV FFAG accelerator, *Proc. 1961 Int. Conf. High-Energy Accelerators* (USAEC, 1961), p. 57.
MURA 1964	MURA staff, The MURA 50-MeV electron accelerator design and construction. 15 papers by various authors, *Rev. Sci. Instrum.* **35**, 1393 (1964).

MURA462   F. T. Cole, High current effects in FFAG accelerators, MURA Report No. 462 (May 1959).

MURA466   J. van Bladel, Image forces in the third MURA model, MURA Report No. 466 (June 1959).

MURA622   T. Edwards, Proton linear accelerator cavity calculations, MURA Report No. 622 (1961), unpublished.

MURA713   D. E. Young, B. Austin, T. W. Edwards, J. E. O'Meara, M. L. Palmer and D. A. Swenson, The design of proton linear accelerators for energies up to 200 MeV, MURA Report No. 713 (July 1, 1965).

MURA714   D. A. Swenson, Application of calculated fields to the study of particle dynamics, *1964 Linear Accelerator Conference Proceedings*, MURA Report No. 714 (1964), p. 328.

Neil 1961   V. K. Neil and A. M. Sessler, Coherent electromagnetic effects in high-current particle accelerators: I. Interaction of a particle beam with an externally driven radio-frequency cavity, *Rev. Sci. Instrum.* **32**(3), 256 (1961).

Neil 1965   V. K. Neil and A. M. Sessler, Longitudinal resistive instabilities of intense coasting beams in particle accelerators, *Rev. Sci. Instrum.* **36**, 429 (1965).

Neilsen 1959A   C. E. Nielsen, A. M. Sessler and K. R. Symon, Longitudinal instabilities in intense relativistic beams, *Int. Conf. High-Energy Accelerators and Instrumentation, CERN, Conf. Proc.* (1959), p. 239 (MURA Report No. 441).

Neilsen 1959B   C. E. Nielsen and A. M. Sessler, Longitudinal space charge effects in particle accelerators, *Rev. Sci. Instrum.* **30**(2), 80 (1959) (MURA Report No. 480).

Ohkawa 1954   T. Ohkawa. Japanese paper.

Ohkawa 1955   T. Ohkawa, FFAG electron cyclotron, *Phys. Rev.* **100**, 1247 (1955).

Ohkawa 1958   T. Ohkawa, Two-beam fixed field alternating gradient accelerator, *Rev. Sci. Instrum.* **29**, 108 (1958) (abstract).

O'Meara 1964   J. E. O'Meara, Summary of Informal Session on Mechanical Engineering Aspects, *Proc. Linear Accelerator Conference MURA* (1964), pp. 616–621.*

---

* The conference proceedings can be found as MURA Report No. 714, or from the Clearinghouse for Federal Scientific and Technical Publications, National Bureau of Standards, US Department of Standards, as *UC-28: Particle Accelerators and High Voltage Machines, TID-4500* (37th edn).

O'Neill 1956   G. K. O'Neill, Storage-ring synchrotrons: Device for high
                  energy physics research, *Phys. Rev.* **102**, 1418 (1956).

Paris 2003   E. Paris, Do you want to build such a machine?":
                  Designing a high-energy proton accelerator for Argonne
                  National Laboratory, Argonne National Laboratory
                  Report ANL/HIST-2 (November 23, 2003).

Parzen 1961   G. Parzen *et al.*, Computer studies and experimental
                  measurements of field perturbations in the MURA
                  two-way electron accelerator, *BNL Accelerator Conf.*
                  (1961), pp. 479–486.

Parzen 1963   G. Parzen and P. L. Morton, Effects of field perturbations
                  in FFAG accelerators, *Dubna Conf. Proc.* (1963),
                  pp. 840–843.

PDG2006   Review of particle physics, *J. Phys.* **33**, 252 (2006).

Pellegrini 1995   C. Pellegrini and A. M. Sessler, *The Development of Colliders*
                  (American Institute of Physics Press, New York, 1995).

Pentz 1961   M. J. Pentz and N. Vogt-Nilsen, A study of orbits in the
                  two-way scaling FFAG synchrotrons, *BNL Accelerator Conf.*
                  (1961), pp. 324–343.

Piccioni 1955   O. Piccioni *et al.*, External beam of the Cosmotron, *Rev.
                  Sci. Instrum.* **26**, 232 (1955).

Pruett 1963A   C. H. Pruett *et al.*, The stacked beam instability in the
                  MURA 50 MeV accelerator, TN-407, MURA Technical
                  Note, 1963 (unpublished).

Pruett 1963B   C. H. Pruett *et al.*, Coherent vertical instabilities in the
                  MURA 50 MeV electron accelerator, *Proc. 1963 Summer
                  Study on Storage Rings, Accelerators, and Experimentation at
                  Super-High Energies* (BNL, 1963), pp. 368–374.

Pruett 1965   C. H. Pruett, F. E. Mills and R. A. Otte, Electronic
                  feedback system for damping the coherent vertical
                  instability in the MURA 50 MeV electron accelerator,
                  Bull. Am. Phys. Soc. **4**, 458 (1965).

Rowe 1965   E. M. Rowe *et al.*, Design of a 200 MeV electron–positron
                  ring, *Proc. Int. Conf. High Energy Accelerators* (Frascati,
                  1965), pp. 279–284 (presented by F. E. Mills).

Rowe 1966   E. M. Rowe *et al.*, Status of the MURA 200 MeV
                  electron–positron storage ring, *Int. Symp. Electron
                  and Positron Storage Rings at Saclay* (1966), III.3,
                  pp. 1–13.

Sands 1959   M. W. Sands, Ultra-high energy synchrotron, MURA
                  Report No. 465 (June 10, 1959).

Sessler 1965   A. M. Sessler, Instabilities of relativistic particle beams, *Proc. Int. Conf. High Energy Accelerators* (Frascati, 1965), pp. 319–330.

Sessler 1966A   E. D. Courant and A. M. Sessler, Transverse coherent resistive instabilities of azimuthally bunched beams in particle accelerators, *Rev. Sci. Instrum.* **37**, 579 (1966).

Sessler 1966B   A. M. Sessler, Summary paper on beam behavior, *Int. Symp. Electron and Positron Storage Rings at Saclay* (1966), IX.1, pp. 1–15.

Sessler 1967   A. M. Sessler and V. Vaccaro, Longitudinal instabilities of azimuthally uniform beams in circular vacuum chambers with walls of arbitrary electrical properties, CERN-67-2 (1967).

Sessler 2007   A. M. Sessler and E. Wilson, *Engines of Discovery: A Century of Particle Accelerators* (World Scientific, Singapore, 2007).

Skaggs 1946   L. S. Skaggs, G. M. Almy, D. W. Kerst and L. H. Lanzi, *Phys. Rev.* **70**, 95 (1946).

Snowdon 1963   S. C. Snowdon, R. S. Christian and G. B. del Castillo, Spiral sector FFAG magnet design and field measurements, *Dubna Conf. Proc.* (1963), pp. 578–589.

Snowdon 1965   S. C. Snowdon *et al.*, Design of a 500 MeV FFAG injector, *Proc. Int. Conf. High Energy Accelerators* (Frascati, 1965), pp. 128–134.

Swenson 1963   D. Swenson, On the threshhold for the coherent vertical instability, TN-423, MURA Technical Note (1963), unpublished.

Swenson 1964A   D. A. Swenson, Application of calculated fields to the study of particle dynamics, *Proc. Linear Accelerator Conference, MURA* (1964), pp. 328–340.

Swenson 1964B   D. A. Swenson, Informal discussion of sparking phenomena, *Proc. Linear Accelerator Conference at MURA* (1964), pp. 606–615.

Symon 1954   K. R. Symon, A strong focussing accelerator with a DC ring magnet, MURA ?, 8/13/1954.

Symon 1955A   K. R. Symon, Fixed field alternating gradient accelerators, *Phys. Rev.* **98**, 1152 (1955) (abstract).

Symon 1955B   K. R. Symon, General theory of orbits in FFAG accelerators, *Phys. Rev.* **100**, 1247 (1955) (abstract).

Symon 1956A   K. R. Symon, D. W. Kerst, L. W. Jones, L. J. Laslett and K. M. Terwilliger. Fixed-field alternating-gradient particle accelerators, *Phys. Rev.* **103**, 1837 (1956).

Symon 1956B	K. R. Symon and A. M. Sessler, Methods of radio frequency acceleration in fixed field accelerators with applications to high current and intersecting beam accelerators, *CERN Symp. Proc.* **I**, 279 (1956).
Symon 1956C	L. J. Laslett and K. R. Symon, Particle orbits in fixed field alternating gradient accelerators, *CERN Symp. Proc.* **1**, 279 (1956).
Symon 1959A	MURA staff, The MURA two-way electron accelerator, *CERN Symp. Proc.* (1959), pp. 71–74.
Symon 1965	K. R. Symon, J. D. Steben and L. J. Laslett, Resonant stability limits for synchrotron oscillations, *Proc. Int. Conf. High Energy Accelerators* (Frascati, 1965), pp. 296–308.
Symon 1991	K. R. Symon, Reflections of the MURA years, *Kent M. Terwilliger Memorial Symposium, AIP Conf. Proc.* **237**, 53 (1991).
Teng 1956	L. C. Teng, *Rev. Sci. Instrum.* **27**, 106 (1956).
Teng 1960	L. C. Teng, Transverse space charge effects, Argonne National Laboratory, Int. Report ANLAD-59 (1960).
Terwilliger 1955A	K. M. Terwilliger, L. W. Jones, F. T. Cole, D. W. Kerst and R. O. Haxby, Small model FFAG betatron I: General description (November APS Meeting, Chicago), *Phys. Rev.* **100**, 1246 (1955) (abstract).
Terwilliger 1955B	K. M. Terwilliger and L. W. Jones, Small model FFAG betatron III: Preliminary results, (November APS Meeting, Chicago), *Phys. Rev.* **100**, 1247 (1955).
Terwilliger 1955C	K. M. Terwilliger, L. W. Jones, K. R. Symon and D. W. Kerst, Application of the fixed field alternating gradient principle to betatrons and cyclotrons, *Phys. Rev.* **98**, 1153 (1955) (abstract).
Terwilliger 1957A	K. M. Terwilliger, L. W. Jones and C. H. Pruett, Beam stacking experiments in an electron model FFAG accelerator, *Rev. Sci. Instrum.* **28**, 987 (1957).
Terwilliger 1957B	K. M. Terwilliger, L. W. Jones, C. H. Pruett and F. T. Cole, Misalignment experiments and theory in a small FFAG betatron, *Bull. Am. Phys. Soc.* **11**(2), 11 (1957).
Terwilliger 1957C	K. M. Terwilliger and L. W. Jones, Radio-frequency with a FFAG accelerator II: Results, *Bull. Am Phys. Soc.* **11**(2), 188 (1957).
Terwilliger 1959	K. M. Terwilliger, Achieving higher beam densities by superposing equilibrium orbits, *CERN Symp. Proc.* (1959), pp. 53–57.

Thomas 1938	L. H. Thomas, The paths of ions in the cyclotron, *Phys. Rev.* **54**, 580 (1938).
Tuck 1950	J. L. Tuck and L. C. Teng, Synchrocyclotron Progress Report III (University of Chicago, Institute of Nuclear Studies, 1950), Chap. 8.
Van Bladel 1960	J. Van Bladel, MURA Report No. 481 (1959); or, IRE *Trans. Microwave Theory and Techniques* MTT-8, 309 (1960).
Veksler 1956	V. I. Veksler, Coherent principles of acceleration of charged particles, *CERN Symp. High Energy Accelerators and Pion Physics* (1956), p. 80.
Waldman 1961	MURA staff, The design of a 100 BeV FFAG accelerator, *Int. Conf. Accelerators* (Brookhaven, 1961), pp. 57–63.
Wallen 1961	MURA staff, Progress on the MURA two-way electron accelerator, *BNL Accelerator Conf.* (1961), pp. 344–351.
Wilkins 1955	J. J. Wilkins, Design of RF resonant cavities for acceleration of protons from 50 to 150 MeV, *P.L.A.C.* **11** (1955).
Wilson 1959	R. R. Wilson, The electron synchrotron, *Handbuch der Physik*, Vol. XLIV, eds. S. Flugge and E. Creutz (Springer-Verlag, Berlin, 1959), p. 176.
Wright 1954	B. T. Wright, Magnetic deflection for the bevatron, *Rev. Sci. Instrum.* **25**, 429 (1954).
Young 1963	D. E. Young, R. S. Christian, C. D. Curtis, T. W. Edwards, F. J. Kriegler, F. E. Mills, P. L. Morton, D. A. Swenson and J. van Bladel, Design studies of proton linear accelerators, *Int. Conf. High Energy Accelerators* (Dubna, USSR, August 21–27, 1963), pp. 454–461.
Young 1964	D. E. Young, Drift tube and parameter selections for linear accelerator structures below 150 MeV, *Proc. Linear Accelerator Conference, MURA* (1964), pp. 177–185.
Young 1966	D. E. Young, C. W. Owen and C. A. Radmer, Preliminary sparking and X-ray phenomena in a 200 Mc linac cavity, MURA Report TN-535 (1965), unpublished; or High field measurements at 200 MHz in conventional proton linear accelerator geometries at 5, 50, and 130 MeV, *Proc. 1966 Linear Accelerator Conference*, Los Alamos Report LA-3609 (1966), pp. 176–182.

Appendix A

GLOSSARY

AdA	Ch. 3	Anello di Annichilazione, the very first electron–positron collider ring, at INFN Frascati.
AEC	Ch. 3	The Atomic Energy Commission, a US government agency; later incorporated into the Department of Energy (DOE).
Alternate Gradient Synchrotron (AGS)	Ch. 4	A 31 GeV proton alternating gradient (strong focusing) synchrotron at Brookhaven National Laboratory, commissioned in 1961.
alternating gradient (AG)	Ch. 2	A method of focusing charged particles in an accelerator (or particle beam), with a sequence of beam elements which are alternately focusing and defocusing; also referred to as strong focusing.
Alvarez linac	Ch. 5	A linac for accelerating heavy particles. It consists of resonant cavities with electrodes (drift tubes) to shield the particles when the accelerating field is in the reverse direction.
Argonne National Laboratory (ANL)	Ch. 2	A DOE research laboratory near Chicago, Illinois, established in 1946.
ARPA	Ch. 5	The Advanced Research Projects Agency of the US Defence Department.
Associated Universities Incorporated (AUI)	Ch. 2	The original contractual organization for Brookhaven National Laboratory.
barn	Ch. 3	A measure of the probability of a reaction, or cross section, expressed as an area; 1 barn = 10^{-24} cm^2.

beam stacking	Ch. 3	Small pulses of particles are accelerated and accumulated to a large equilibrium distribution of particles circulating at constant energy in a synchrotron or storage ring.
betatron	Ch. 1	An electron circular accelerator that utilizes the electric field induced by a changing magnetic field to accelerate the electrons.
betatron oscillation	Ch. 3	The bounded transverse oscillatory motion of particles about the design (equilibrium) orbit in a circular accelerator.
BeV (GeV)	Ch. 4	A billion electron volts (10^9 eV) of energy.
Bevatron	Ch. 2	A 6 GeV proton synchrotron at LBNL, commissioned in 1954.
boson	Ch. 5	A particle of force in the Standard Model. At the quantum level, each force in nature is manifested in a particle or combination of particles.
Brookhaven National Laboratory (BNL)	Ch. 2	A basic research laboratory on Long Island, New York, funded by the DOE.
bucket (rf)	Ch. 3	The area in rf phase space — rf phase versus particle energy — within which rf oscillations are stable.
c.m.		Center of mass; the coordinate system in which the vector sum of the momenta of a set of particles is zero.
Cambridge Electron Accelerator (CEA)	Ch. 3	A Harvard–MIT 6 GeV electron accelerator.
Calutron	Ch. 4	A cyclotron-like device for separating isotopes of uranium.
European Organization for Nuclear Research (CERN)	Ch. 2	The European laboratory for particle physics in Geneva, Switzerland, and neighboring France; established in 1954. The world's largest particle physics laboratory.
chromaticity	Ch. 4	The dependence of focusing on particle momentum that results in a dependence of

		betatron oscillation tune on momentum, analogous to optical chromatic aberration.
circumference factor	Ch. 3	The circumference of an accelerator ring divided by 2π times the minimum radius of curvature.
Cockcroft–Walton	Ch. 2	A type of electrostatic accelerator, dating from 1932.
collider	Ch. 3	An accelerator system in which the target for a particle beam is a second particle beam moving in the opposite direction.
Cosmotron	Ch. 2	A 3 GeV proton synchrotron at Brookhaven National Laboratory, commissioned in 1952.
cyclotron	Ch. 2	A circular proton accelerator with a dc magnet, dating from the early 1930s.
debuncher	Ch. 5	A device for reducing the energy spread of a particle beam.
Department of Terrestrial Magnetism	Ch. 2	Part of the Carnegie Institute in Washington.
Deutsches Elektronen Synchrotron, "German Electron Synchrotron" (DESY)	Ch. 3	The high-energy laboratory in Hamburg, Germany.
dipole	Ch. 3	A magnet of spatially uniform field strength that bends charged particles in a circular arc.
Department of Energy (DOE)		The US Department of Energy (earlier, AEC).
drift tube linac (DTL)	Ch. 5	A type of linear accelerator most suitable for the acceleration of heavy particles (e.g., protons); see also Alvarez linac.
Joint Institute for Nuclear Research (Dubna)	Ch. 2	A laboratory (north of Moscow) of countries associated with the Soviet Union (now Russia).
electrostatic accelerator	Ch. 3	A type of accelerator where a dc electrostatic voltage provides the acceleration.

Electron Machine with Many Applications (EMMA)	Ch. 3	An experiment at Daresbury Laboratory, England, on nonscaling FFAGs.
emittance	Ch. 4	The phase space area occupied by the beam.
eV	Ch. 3	A unit of energy equal to that of an electron accelerated across a potential of 1 V, equivalent to 1.6×10^{-19} J.
ferroelectric	Ch. 3	A material with a permanent electric field.
Fermilab (FNAL), formerly NAL	Ch. 1	Fermi National Accelerator Laboratory. A particle physics research laboratory in Batavia, Illinois; near Chicago; funded by DOE.
FFAG	Ch. 3	Fixed Field Alternating Gradient. An annular accelerator with a magnetic field fixed in time, with azimuthally varying fields to achieve strong focusing and a radial gradient such that equilibrium orbits over a wide range of particle energies are contained.
focusing	Ch. 3	In an accelerator, the means by which particles away from the equilibrium (ideal, stable) orbit are moved back toward (or across) it as they are accelerated and/or circulated.
gantry	Ch. 3	A multiangle beam delivery system for therapy.
gradient	Ch. 2	The spatial variation of a magnetic field.
gauss	Ch. 3	A unit of magnetic field strength.
GeV	Ch. 3	A giga-electron-volt (10^9 eV) of energy (also, earlier, BeV).
hadron	Ch. 2	An elementary particle of heavy mass which experiences the "strong interaction," such as a proton, neutron, pion, kaon or hyperon.
Hamiltonian	Ch. 3	The energy of the system, expressed in canonical coordinates, so that the equations of motion can be readily obtained from it.
heliarc	Ch. 3	An aluminum welding technique using helium gas to minimize oxidation of the weld.

HERA	Ch. 3	Hadron–Electron Ring Accelerator: the proton–electron collider at DESY.
hertz (Hz)		Oscillation frequency in cycles per second.
High-Energy Physics Laboratory (HEPL)	Ch. 3	A laboratory of Stanford University.
hyperon	Ch. 2	A baryon (nuclear hadron) heavier than the proton or neutron, composed of three quarks, including a "strange" or "charm" quark.
ILLIAC	Ch. 3	An early (1950s) digital computer at the Illinois Automatic Computer, University of Illinois.
Institute of Nuclear Physics (INP)	Ch. 3	A laboratory in Novosibirsk (Siberia). Now called the Budker Institute of Nuclear Physics (BINP).
Intersecting Storage Rings (ISR)	Ch. 3	The world's first proton–proton storage ring, at CERN.
Istituto Nazionale di Fisica Nucleare (INFN)	Ch. 3	The nuclear agency in Italy, with laboratories throughout the country.
K of a cyclotron	Ch. 3	The maximum proton energy in MeV that can be accelerated in a particular cyclotron.
k	Ch. 3	A measure of the magnetic field gradient (and nonlinear terms also). The magnetic field increases as r^k.
kaon	Ch. 2	k meson; a heavy meson, containing a strange quark or antiquark, with mass between that of a pion and a proton.
keV	Ch. 3	A kilo-electron-volt (10^3 eV) of energy.
Kyoto University Research Reactor Institute (KURRI)	Ch. 3	The reactor laboratory of Kyoto University, Japan.
Landau damping	Ch. 3	Damping of coherent oscillations due to dependence of frequency on energy.
Large Hadron Collider (LHC)	Ch. 3	CERN's very large proton–proton collider, completed in 2008.

Lawrence Berkeley National Laboratory (LBNL), formerly LBL and LRL	Ch. 3	A research laboratory at the University of California at Berkeley; funded by DOE.
lepton	Ch. 3	A weakly interacting elementary particle such as an electron, muon, or neutrino.
LHC	Ch. 3	The Large Hadron Collider at CERN, a 27 km circumference proton–proton colliding beam accelerator/storage ring with beams of 7 TeV, to produce 14 TeV c.m.; to be commissioned in 2008.
linac	Ch. 2	A linear accelerator for charged particles (protons or electrons) in which a series of electrodes are arranged so that, when a voltage is applied at the proper radio frequency, the particles passing through them receive, in phase, successive increments of energy.
Liouville theorem	Ch. 3	A physics theorem concerning the conservation of the phase space occupied by an ensemble of particles, e.g., in accelerator orbits.
luminosity	Ch. 3	The interaction rate per unit of interaction cross section in a colliding beam accelerator or storage ring.
magnetohydro-dynamics (MHD)	Ch. 5	The study of the behavior of an electrically conducting fluid moving in a magnetic field.
Mark I	Ch. 3	The original FFAG accelerator geometry.
Mark Ia	Ch. 3	In this FFAG accelerator the field strength has a greater magnitude in the radially focusing segments, and adjacent segments have reverse polarity but equal length, hence providing a net positive curvature.
Mark Ib	Ch. 3	An FFAG accelerator in which the field strength has the same magnitude in all segments, but adjacent segments have reverse polarity; the radially focusing magnets are longer than the radially

		defocusing magnets, to provide net positive curvature.
Mark II	Ch. 3	A nonscaling FFAG where the circumference factor is bad at injection and the length of the reverse field segment is reduced for the high-energy orbits in order to reduce the circumference factor at maximum energy.
Mark III	Ch. 3	Similar to Mark Ib, but with a better circumference factor by operating on an island of stability in the "necktie diagram" but where vertical focusing is very weak and radial focusing is much stronger.
Mark IV	Ch. 3	A grossly nonscaling FFAG design in which the magnetic field direction is the same in all magnets and the field gradients alternate (to achieve a better circumference factor).
Mark V	Ch. 3	An FFAG which has spiral ridge magnetic pole structure in which the field in all magnets is in the same direction, but the edges of the spiral magnets provide alternating positive and negative focusing in order to retain stability in both dimensions.
meson	Ch. 2	An unstable particle with masses between the electron and proton, e.g. a pion (pi meson), a kaon (k meson), or a muon (mu meson).
MeV	Ch. 3	A million electron volts (10^6 eV) of energy.
midplane	Ch. 2	The central plane of an accelerator.
Midwestern Universities Research Association (MURA)	Ch. 1	An association created for the purpose of bringing a high-energy accelerator to the Midwest.
microampere (μA)	Ch. 3	A millionth of an ampere (10^{-6} A).
microsecond (μs)	Ch. 3	A millionth of a second (10^{-6} s).
milliamp (mA)	Ch. 4	A thousandth of an ampere (10^{-3} A).
millisecond (ms)		A thousandth of a second (10^{-3} s).

muon	Ch. 2	μ meson (mu meson); a weakly interacting charged particle of mass 106 MeV/c^2, or ~207 times the electron mass, and with a decay lifetime of 2.2 μs.
multipole	Ch. 5	A field produced by alternating north and south magnetic poles as in quadrupole (4) or sextapole (6).
nanosecond (ns)	Ch. 5	A billionth of a second (10^{-9} s).
NbSn	Ch. 5	A niobium–tin alloy that becomes superconducting at a few degrees Kelvin.
NbTi	Ch. 5	A niobium–titanium alloy that becomes superconducting at a few degrees Kelvin.
neutrino		A weakly interacting elementary particle of very small mass and no electric charge; there are electron, muon, and tau neutrinos.
neutron		A stable nuclear particle of zero charge and a mass 1836 times the mass of an electron, or 938 MeV/c^2.
National Science Foundation (NSF)	Ch. 4	The US agency which supports basic research.
nucleon		A neutron or proton.
octupole	Ch. 4	A magnetic element with eightfold azimuthal symmetry; four north and four south magnetic poles, alternating.
Office of Naval Research (ONR)	Ch. 5	The US Navy office that provided funds for research.
Office of Scientific Research (OSR)	Ch. 5	US Air Force Office of Scientific Research.
Paul Scherrer Institute (PSI)	Ch. 3	A major laboratory in Switzerland.
phase space	Ch.3	The six-dimensional space whose coordinates are the canonical variables of position and momenta of a particle. The space considered is often 2 dimensional (x and p_x)

		or (energy/frequency and phase) but must be six-dimensional to describe completely describe a particle's motion.
photodesorption	Ch. 5	The removal of molecules from a surface by electromagnetic radiation (ultraviolet light or X-rays, for example).
pion	Ch. 2	π meson (pi meson); a strongly interacting charged or neutral particle of a mass ~280 times that of an electron, or 140 MeV/c^2 (charged pion); 135 MeV/c^2 (neutral pion), and a decay lifetime of 2.6×10^{-8} seconds (charged pion).
proton	Ch. 3	A stable nuclear particle of charge +1 and a mass 1836 times that of an electron, or 938 MeV/c^2.
Proton Synchrotron (PS)	Ch. 3	A 28 GeV strong-focusing synchrotron at CERN, commissioned in 1959.
Physical Sciences Laboratory (PSL)	Ch. 5	The Physical Sciences Laboratory of the University of Wisconsin.
quadrupole	Ch. 2	A magnetic focusing element with fourfold azimuthal symmetry; two north and two south magnetic poles, alternating.
quark	Ch. 5	The particles in the Standard Model that make up protons, neutrons, and a variety of lesser-known particles.
rf	Ch. 3	The frequency of an accelerating electromagnetic field, in the range of radio frequencies.
Radio Frequency Quadrupole (RFQ)	Ch. 2	An element that provides transverse focusing in an rf linear accelerator, also, an accelerating structure incorporating this principle.
relativistic	Ch. 3	Motion with speed approaching that of light.
Relativistic Heavy Ion Collider (RHIC)	Ch. 3	A BNL accelerator/colliding beam facility for colliding heavy ions of energies about 100 GeV per nucleon; 200 GeV per

		nucleon pair c.m. energy; commissioned in 2001.
revolution frequency	Ch. 3	The number of times per second a particle travels around a cyclic accelerator.
RIKEN	Ch. 3	The Institute of Physical and Chemical Research, Japan.
rhumbatron	Ch. 4	A type of radio frequency cavity.
Runge–Kutta	Ch. 3	A step-by-step integration method.
scaling	Ch. 3	A type of accelerator where equilibrium orbits of accelerated particles at different energies are identical except for a radial scale factor.
sextupole	Ch. 4	A magnetic element with sixfold azimuthal symmetry; three north and three south magnetic poles, alternating.
space charge	Ch. 4	The electric charge of a group of particles.
SPEAR	Ch. 3	An early e^+–e^- colliding beam storage ring at SLAC.
solenoid	Ch. 4	A cylinder wrapped with current-carrying wires.
stacking	Ch. 4	A method of accumulating beam while closely preserving phase space density.
Super-Proton Synchrotron (SPS)		A 400 GeV proton synchrotron at CERN; commissioned in 1975, later utilized for proton–antiproton collisions.
Superconducting Super Collider (SSC)	Ch. 3	A 1983 US-proposed 20 GeV accelerator and proton–proton collider, to produce 40 TeV c.m. It was canceled in 1992 and never built.
Stanford Linear Accelerator Center (SLAC)	Ch. 4	A DOE-supported research laboratory at Stanford University in Palo Alto, California.
stochastic cooling	Ch. 3	A method of reducing the phase space occupied by a circulating beam, based upon detecting the fluctuations and supplying a damping voltage signal to reduce them.

storage ring	Ch. 3	A dc magnet ring like that of a synchrotron in which particles are not accelerated, but are stored at constant energy, circulating around the ring in stable orbits.
strong focusing		Also known as alternating gradient (AG) focusing.
superferric	Ch. 5	The use of iron to help shape the magnetic field in a superconducting magnet.
symplectic	Ch. 3	Motion described by a Hamiltonian.
Synchro-Phasotron	Ch. 2	A 10 GeV synchrotron in Dubna.
synchrotron	Ch. 2	A circular accelerator in which particles are kept in a closed orbit by a magnetic guide field that rises as they are accelerated by an applied rf system, which is frequency-modulated to match the revolution frequency of the particles as they are accelerated.
Tantalus	Ch. 5	A synchrotron radiation ring at PSL.
tesla (T)	Ch. 5	A unit of magnetic field strength, equal to 10^4 gauss.
Tevatron Collider	Ch. 3	The proton–antiproton collider at Fermilab.
thyratron	Ch. 3	A radio tube capable of producing high-voltage pulses.
torr	Ch. 3	A measure of gas pressure, corresponding to one mm Hg, or 1/760 of a standard atmosphere.
transition energy	Ch. 3	That particle energy at which the revolution frequency in an accelerator does not change incrementally with energy. Below the transition energy the frequency increases with energy; above the transition energy the frequency decreases with energy.
National Laboratory for Particle and Nuclear Physics (TRIUMF)	Ch. 3	The cyclotron laboratory in Vancouver, Canada.

Universities Research Association (URA)	Ch. 5	The former contracting organization for Fermilab.
vacuum chamber		In an accelerator (or beam line), an enclosure in which the particle beam is contained, from which the air has been extracted.
Variable Energy Cyclotron Centre	Ch. 3	A laboratory of the Department of Atomic Energy in Kolkata, India.
water bag model	Ch. 4	A model used for theoretical calculation of space charge effects.
Van de Graaff accelerator	Chs. 2 & 4	An early type of linear, dc electrostatic accelerator.
X-rays	Ch. 2	Electromagnetic radiation of energy in the range of ~1 keV to ~ 1 MeV.
Zero Gradient Synchrotron (ZGS)	Ch. 5	A 12 GeV proton synchrotron at Argonne National Laboratory; commissioned in 1962.
μf	Ch. 3	Microfarad; a unit of electrical capacitance.

MURA REPORTS

The following reports are found in the Fermilab library. The list is complete except for a few reports pertaining to the currently obsolete computing system in use at MURA before 1967. A complete file of all the MURA reports can be found in the archives of the Physical Sciences Laboratory of the University of Wisconsin. Also, the reports can be found at the website www-spires.fnal.gov

MURA-a
1959 MURA summer study reports on design and utilization of high-energy accelerators

MURA-b
The MURA 50-MeV electron accelerator: Design and construction

MURA-c; Cole, F. T.
A guide to particle accelerators

MURA-d; Rowe, E. M.
Design of a 200-MeV electron–positron storage ring

MURA-e; Rowe, E. M.
Status of the MURA 200-MeV electron–positron storage ring

MURA-f; Snowdon, S. C.
A proposal for advanced studies in particle accelerators

MURA-001; Brueckner, K. A.; Watson, K. M.
Research with high-energy accelerators

MURA-002; Kerst, D. W.
An estimate of effects of nonlinear restoring forces for avoiding resonance, 1953

MURA-003; Courant, Ed
Linear coupling between vertical and horizontal oscillation, 1953

MURA-004; Cole, F. T.; Rohrlick, F.
A perturbation treatment of nonlinear restoring forces, 1953

MURA-005; Wright, S. C.
Adiabatic damping of large phase oscillation, 1953

MURA-006; Jones, L. W.; Laslett, L. J.
A study of the feasibility of a multi-BeV circular electron accelerator, 1953

MURA-007; Terwilliger, K. M.
Magnet aperture as a function of N, 1953

MURA-008; Kerst, D. W.
An ion pipe: an extreme form of AG magnet, 1953

MURA-009; Crane, H. R.
Proposal for a high intensity accelerator, 1953

MURA-010; Kerst, D. W.
Magnet power supply, 1953

MURA-011; Powell, J. L.
Nonlinearities in the AG synchroton, 1953

MURA-012; Powell, J. L.
Note on discontinuity in field gradient, 1953

MURA-013; Jones, L. W.
A note on the amplitude of betatron oscillations, 1953

MURA-014; Laslett, L. J.
Discussion of space charge effects in the alternate gradient synchrotron, 1954

MURA-015
Key to digital computer photographs, 1954

MURA-016; Kerst, D. W.
Approximate calculation on nonlinear lock-in at sigma = pi, 1954

MURA-017; Kerst, D. W.
Characteristics of nonlinear lock-in caused by field inhomogeneity, 1954

MURA-018; Laslett, L. J.
Possible instability from momentum errors in the AGS, 1954

MURA-019; Laslett, L. J.
Equations of motion in the AGS, 1954

MURA-020; Jones, L. W.; Terwilliger, K. M.
An electromechanical analogue for the study of strong focusing synchrotron betatron orbits, 1954

MURA-021; Powell, J. L.
Nonlinearities in AG synchrotron, 1954

MURA-022; Kerst, D. W.; Elfe, T. B.
Paths of particles in cicularly symmetric fringing fields, Stormer method, 1954

MURA-023; Laslett, L. J.
Damping of oscillations: requisite energy tolerance at injection coherent radiation, 1954

MURA-024; Laslett, L. J.
Requisite energy tolerance at injection, 1954

MURA-025; Symon, K. R.
A smooth approximation to the alternating gradient orbit equations, 1954

MURA-026; Symon, K. R.
Smooth solution to one-dimensional AG orbits with cubic forces, 1954

MURA-027; Livingood, J. J.
A suggestion to eliminate space charge in a proton synchrotron, 1954

MURA-028; Jones, L. W.
Possible design parameters for a 10-BeV high repetition rate proton alternating gradient synchrotron, 1954

MURA-029; Jones, L. W.
Values of alpha off the diagonal of the stability diagram, 1954

MURA-030; Symon, K. R.
An adiabatic theorem for motions which exhibit invariant phase curves, 1954

MURA-031; Symon, K. R.
An alternative derivation of the formulas for the smooth approximation, 1954

MURA-032; Symon, K. R.
A strong focussing accelerator with a DC ring magnet (preliminary report for internal circulation only), 1954

MURA-033; Laslett, L. J.
Retabulation of space charge effects in the AGS, 1954

MURA-034; Kerst, D. W.
Report for the National Science Foundation on the studies of the Midwestern Universities Research Association

MURA-035; Crosbie, E. A.; Hammermesh, M.
Coupling of betatron and phase oscillations in a synchrotron, 1954

MURA-036; Kerst, D. W.
Comments on machines for the energy range 2 BeV to 10 BeV

MURA-037; Haxby, R. O.
Report on magnetic measurements at Brookhaven

MURA-038; Sachs, R. G.
On the application of very high-energy machines

MURA-039; Vogt-Nilsen, N.
The influence of radiation on oscillating electron orbits in an ideal betatron, 1954

MURA-040; Laslett, L. J.
Concerning the attainment of stable orbits with negative momentum compaction, 1954

MURA-041; Terwilliger, K. M.; Jones, L. W.
Application of FFAG principle to betatron acceleration, 1953

MURA-042; Kerst, D. W.; Jones, L. W.
A fixed field alternating gradient accelerator with spirally ridged poles, 1954

MURA-043; Symon, K. R.
The FFAG synchrotron mark I, 1954

MURA-044; Jones, L. W.
Patch operation of a fixed field alternating gradient accelerator, Mark III, with comments on momentum compaction and amplitudes of oscillation, 1954

MURA-045; Jones, L. W.; Kerst, D. W.; Laslett, L. J.
A Fixed Field Alternating Gradient synchrotron with a small circumference factor, 1954

MURA-046; Symon, K. R.
FFAG with spiral poles, smooth approximation, 1954

MURA-047-1; Jauch, J. M.
The stability of orbits in a nonlinear AG synchrotron. 1: Phase functions for canonical transformations, 1954

MURA-047-2; Jauch, J. M.
Stability theory. 2: Some special canonical transformations in two-dimensional phase space, 1954

MURA-048; Jones, L. W.
Edge effects in FFAG, MARK-II, 1954

MURA-049-1; Powell, J. L.; Wright, R. S.
Nonlinearities in A. G. synchrotrons. Part I: A Survey of results obtained by use of fast digital computer, 1955

MURA-049-2; Powell, J. L.; Wright, R. S.
Nonlinearities in A. G. synchrotrons. Part II: Approximate transformations of the phase plane, 1955

MURA-050; Kerst, D. W.
Negative momentum compaction in conventional A. G. synchrotrons, 1955

MURA-051; Laslett, L. J.
Estimate of possible spatial variation of N

MURA-052; Jones, L. W.; Terwilliger, K. M.
FFAG Mark Ib formation including edges and straight sections, 1955

MURA-053; Jones, L. W.; Terwilliger, K. M.
A Mark Ib betatron design, 1955

MURA-054; Akeley, E. S.
Expansions which express the magnetic field on either side of a plane surface in terms of the magnetic field on the surface, and their application to the Mark V FFAG accelerator, 1955

MURA-055; Laslett, L. J.
Approximation of Eigenvalues and Eigenfunctions, by variational methods, 1955

MURA-056; Laslett, L. J.
Paraxial lens action in electric ion optics or geometrical optics, 1955

MURA-057; Jones, L. W.
Injection into FFAG accelerators, 1955

MURA-058; Jones, L. W.
A special case of Mark V parallel sloped bars

MURA-059; Terwilliger, K. M.
Vertical aperture and field fluter with slotted pole pieces, 1955

MURA-060; Cole, F. T.
Perturbation theory of AG motion with nonlinear restoring forces

MURA-061; Akeley, E. S.
A method for determining the magnetic field in a region in terms of its values on a plane surface, 1955

MURA-062; Kerst, D. W.
High-energy FFAG ring magnet with spirally ridged field, 1955

MURA-063; Symon, K. R.
Contributions to the unified theory of FFAG fields, 1955

MURA-064; Kerst, D. W.
Constant frequency cyclotrons with spirally ridged poles, 1955

MURA-065; Jones, L. W.
A special case of MARK II with no reverse field, 1955

MURA-066; Kerst, D. W.
Gap in a spirally ridged pole, 1955

MURA-067; Cole, F. T.; Kerst, D. W.
Detailed calculations of a small model FFAG Mark Ib accelerator. Parts A and B, 1955

MURA-068; Elfe, T. B.; Kerst, D. W.
Investigation of the effect of position of straight sections with application to the Mark V spirally ridged accelerator

MURA-069; Francis, N. C.
Radial and phase motion in a synchrotron

MURA-070; Akeley, E. S.
The vector potential of the magnetic field in the Mark V accelerator, 1955

MURA-071; Sessler, A. M.
Determination of sigma in the model FFAG Mark Ib accelerator, 1955

MURA-072; Kerst, D. W.
Distributions of straight sections in Mark V

MURA-073; Vogt-Nilsen, N.
A vector potential expansion and the corresponding equations of motion for a Mark V FFAG accelerator, 1955

MURA-074; Jones, L. W.
Injection into the Mark Ib FFAG model, 1955

MURA-075; Laslett, L. J.
Character of particle motion in the Mark V FFAG accelerator, 1955

MURA-076; Sessler, A. M.
The construction of approximate phase plane transformations for AG synchrotrons, I, 1955

MURA-076-a; Sessler, A. M.
The construction of approximate phase plane transformations for AG synchrotrons, II, 1955

MURA-077; Judd, D. L.
Analytical approximation in Mark V scalloped orbits and to radial betatron oscillations about them, 1955

MURA-078; Judd, D. L.
Nonlinear terms in Mark V radial betatron equation, 1955

MURA-079; Sessler, A. M.
Study of the effects of displaced sectors with approximate phase plane transformations, 1955

MURA-080; Powell, J. L.
Mark V FFAG: Equations of motion for Illiac computation, 1955

MURA-081; Kerst, D. W.
The possibility of high intensities from FFAG accelerators providing a means of increasing the energy, 1955

MURA-082; Cole, F. T.; Haxby, R. O.; Terwilliger, K. M.
Design parameters for an 8 sector FFAG MARK-1B model, 1955

MURA-083; Wimpress, R.; King, R. F.; Laslett, L. J.
The electrostatic field of the kicker electrodes in the BNL electron analog, 1955

MURA-084; Okhawa, T.
An application of FFAG principle to the phase oscillation in synchrotron, 1955

MURA-085; Fosdick, L. D.
Investigations of f-imperfections in an alternating gradient synchrotron with nonlinear forces, 1955

MURA-086; Ohkawa, T.
The effects of the straight sections in Mark V, 1955

MURA-087; Snyder, J. N.
Properties of the RF acceleration program, 1955

MURA-088; Kerst, D. W.
Some estimates of properties of intersecting beam accelerators, 1955

MURA-089; Laslett, L. J.
Stable and unstable periodic orbits in certain Mark V accelerators as obtained by the Ridge Runner computational program, 1955

MURA-090; Sessler, A. M.
Half sector phase plane transformations for an AG synchrotron, 1955

MURA-091; Ohkawa, T.
Indirect acceleration

MURA-092; Sessler, A. M.
Potentials and fields in a scaling accelerator, 1955

MURA-093; Ohkawa, T.
Comments on several modifications of the magnetic field of spiral sector magnets

MURA-094; Laslett, L. J.
Interpolation formulas for a two-dimensional net, 1956

MURA-095; Cole, F. T.
Mark-5 FFAG expanded equations of motion, 1956

MURA-096; Cole, F. T.
Equilibrium orbit in Mark-1 FFAG, 1956

MURA-097; Laslett, L. J.
Proposal to refine the determination of the characteristic exponent in a Hills equation pertaining to the Mark-5 FFAG, 1956

MURA-098; Sessler, A. M.
Acceleration schemes, 1956

MURA-099; Laslett, L. J.
Proposed method for determining Mark V trajectories by aid of grid storage, 1956

MURA-100; Ohkawa, T.
Mark V scalloped motion in the axial direction

MURA-101; Jones, L. W.; Terwilliger, K. M.
A note on the accelerated beam obtained in the Michigan radial sector FFAG electron model, 1956

MURA-102; Crane, H. R.
Maintaining the geometrical alignment of a large accelerator, 1956

MURA-103; Sessler, A. M.
Memo on two-dimensional algebraic transformations which correspond to Hamiltonian systems, 1956

MURA-104; Jones, L. W.; Terwilliger, K. M.
A small model fixed field alternating gradient radial sector accelerator, 1956

MURA-105; Laslett, L. J.
Application of Walkinshaw's equation to the 2 sigma(y) = sigma(x) resonance, 1956

MURA-106; Symon, K. R.; Sessler, K. M.
Methods of radio frequency acceleration in fixed field accelerators with applications to high current and intersecting beam accelerators, 1956

MURA-107; Laslett, L. J.; Sessler, A. M.
The y-stability limit in spiral sector accelerators, 1956

MURA-108; Laslett, L. J.; Symon, K. R.
Particle orbits in fixed field alternating gradient accelerators, 1956

MURA-109; Krest, D. W.; Symon, K. R.; Terwilliger, K. M.
Fixed field alternating gradient particle accelerators, 1956

MURA-110; Lichtenberg, D. B.; Newton, R. G.; Ross, M. H.
Intersecting beam accelerator with storage ring, 1956

MURA-111; Kerst, D. W.
Properties of an intersecting beam accelerating system, 1956

MURA-112; Kerst, D. W.
Spiral sector magnets, 1956

MURA-113; Jones, L. W.
Concentric and eccentric colliding beams geometries, 1956

MURA-114; Sessler, A. M.
Proposed digital computer program to study the coupling of radial betatron oscillations, 1956

MURA-115; Sessler, A. M.
A proposed possible extension of the digital computer program for studying foils and simple RF gaps, 1956

MURA-116; Johnston, L. H.; Schuldt, S.
Adiabatic damping of large amplitude phase oscillations in a linear accelerator

MURA-117; Akeley, E. S.
The magnetic field and the equations of motion in spherical coordinates for the Mark V FFAG accelerator, 1956

MURA-118; Vogt-Nilsen, N.
Expansions of the characteristic exponents and the Floquet solutions for the linear homogeneous second order differential equation with periodic coefficients, 1956

MURA-119; Laslett, L. J.
Analysis and computation of magnetic fields arising from two-dimensional pole configurations of interest in spiral sector accelerators, 1956

MURA-120; Laslett, L. J.; Sessler, A. M.
The x-stability limit in spiral sector accelerators, 1956

MURA-121; Laslett, L. J.
Calculations concerning particle motion in spirally ridged and separated sector FFAG accelerators, 1956

MURA-122; Weinberg, E.
Approximate calculation of the potential at the boundary of a current carrying rectangular coil, at the corner of a pole, 1956

MURA-123; Vogt-Nilsen, N.
A short survey of digital computer results for radial motion in FFAG Mark V spiral ridge accelerators, 1956

MURA-124; Ohkawa, T.
A scaled radial sector FFAG for intersecting beams, 1956

MURA-125; Sessler, A. M.
On the non-Liouvillian character of foils, 1956

MURA-126; Lichtenberg, D. B.; Stehle, P.; Symon, K. R.
Modification of Liouville's theorem required by the presence of dissipative forces, 1956

MURA-127; Cole, F. T.
Tuning of vertical oscillations in separated sector Mark V by twisting magnets, 1956

MURA-128; Sessler, A. M.
The x-stability limit in large f structures, 1956

MURA-129; Parzen, G.
Nonlinear resonances in radial motion, 1956

MURA-130; Kerst, D. W.
A betatron beam stacking process and asynchronous acceleration without a large betatron core, 1956

MURA-131; Laslett, L. J.
Fixed field alternating gradient accelerators, 1956

MURA-132; Lichtenberg, D. B.
Gas scattering in fixed field proton accelerators, 1956

MURA-133; Terwilliger, K. M.
Radio frequency knockout of stacked beams, 1956

MURA-134; Jones, L. W.
The Okhawa intersecting beam machine, 1956

MURA-135; Laslett, L. J.
Results of miscellaneous two-dimensional potential computations, involving pole face currents, of possible interest in FFAG magnet design, 1956

MURA-136; Symon, K. R.
Fundamental limitations on the performance of an adiabatic accelerator, 1956

MURA-137; Lichtenberg, D. B.
A simple method for RF acceleration in a fixed field machine, 1956

MURA-138; Jones, L. W.
Effects of bumps when magnet edges are important, 1956

MURA-139; Laslett, L. J.
Axial amplitude limitations effected by sigma(x) + 2 sigma(y) = 2 pi, 1956

MURA-140; Ohkawa, T.
Beam loading on an RF cavity and an acceleration scheme by velocity modulation, 1956

MURA-141; Jones, L. W.
Long straight sections in separated spiral sector accelerators, 1956

MURA-145; Snyder, J. N.
sigma(1), sigma(2), psi(1), psi(2), eta(2)/eta(1), eta-prime(2)/eta-prime(1) tables, 1956

MURA-147; Belford, G.
Table pertaining to solutions of a Hill equation, 1956

MURA-200; Parzen, G.
Nonlinear resonances in alternating gradient accelerators, 1956

MURA-201; Denman, H. H.
Numerical techniques, 1956

MURA-202; Cole, F. T.; Mewier, H. K.
Ex post facto betatron frequency calculations for the Michigan radial sector FFAG betatron, 1956

MURA-203; Cole, F. T.; Jones, L. W.
Misalignments in the Michigan radial sector FFAG accelerator, 1956

MURA-204; Stump, R.; Pavlat, J.
Electrical characteristics of a model mechanically tuned radio frequency cavity for a multi-BeV synchrotron, 1956

MURA-205; Laslett, L. J.
Development of algorithms for Forocyl potential program, 1956

MURA-206; Laslett, L. J.
Remarks on the invariant quadratic forms pertaining to motion characterized by a linear differential equation with periodic coefficient, 1956

MURA-207; Chang, T.-S.
Energy of colliding proton beams and angular distribution of mesons produced, 1956

MURA-208; Lichtenberg, D. B.
Energy dependence of space charge forces in an accelerator beam, 1956

MURA-209; Ohkawa, T.
Acceleration by fixed frequency cavities independent of the rotational frequency of particles, 1956

MURA-210; Van Bladel, J.
The reaction of a cavity on the beam current, 1956

MURA-211; Laslett, L. J.
Use of a scalar potential in two-dimensional magnetostatic computations with distributed currents, 1956

MURA-212; Jones, L. W.; Pruett, C. H.; Terwilliger, K. M.
Effects of resonances in the radial sector FFAG model, 1956

MURA-213; Laslett, L. J.
Computational checks of particle motion in the Illinois separated sector FFAG model, without perturbations, 1957

MURA-214; Ohkawa, T.
Preliminary considerations of the acceleration of particles by reflection of electromagnetic waves, 1957

MURA-215; Kerst, D. W.
Suppression of betatron oscillation excitation (RF knockout) by the RF accelerating system of a fixed field accelerator, 1957

MURA-216; Kerst, D. W.
Report on visit to Machlett Laboratories, Inc., 1957

MURA-217; Parzen, G.
Coupled nonlinear resonances in alternating gradient accelerators, 1957

MURA-218; Laslett, L. J.
Stable equilibrium orbit in several large scale spiral sector FFAG accelerators, 1957

MURA-219; Cole, F.T.; Haxby, R. O.; Terwilliger, K. M.
An electron model Fixed Field Alternating Gradient accelerator, 1957

MURA-220; Van Bladel, J.
Rough calculations on flange coupled cavities, 1957

MURA-221; Laslett, L. J.; Snyder, J. N.
FOROCYL program (program 13) including FORMERGE (program 50), 1957

MURA-222; Laslett, L. J.
FORMESH program (program 26), 1957

MURA-223; Cole, F. T.; Snyder, J. N.
SCOFFLAW program (program 30), 1957

MURA-224; Symon, K. R.; Joyce, C. A.
JOYBUCKETS (program 32), 1957

MURA-225; Mills, F. E.; Snyder, J. N.
TTT (programme 39): A programme for RF study, 1957

MURA-226; Fosdick, L. D.
Well tempered five (program 46), 1957

MURA-227; Snyder, J. N.
FORMERGE (program 50), 1957

MURA-228; Snyder, J. N.
FORANAL (program 52): A program for Fourier analysis, 1957

MURA-229; Snyder, J. N.
ATEMESH (program 53), 1957

MURA-230; Snyder, J. N.
EQUICYL (program 54), 1957

MURA-231; Snyder, J. N.
TTITLE = Fixed point search, rotation number, and invariant coefficient programmes:
FORFIX Point (programme 55), FORFIX Point with FUMBLEBUMPS (programme
66), FORFIX Point with GRUMBLEBUMPS (programme 67), 1957

MURA-232; Snyder, J. N.
SIXTEEN MESH (program 56), 1957

MURA-233; Storm, M. R.
Algytee (program 58), 1957

MURA-234; Snyder, J. N.
FORMESH FUMBLEBUMPS (program 60), 1957

MURA-235; Snyder, J. N.
FORMESH with MUMBLEBUMPS (program 61), 1957

MURA-236; Snyder, J. N.
FORMESH with GRUMBLEBUMPS (program 62), 1957

MURA-237; Snyder, J. N.
DUCK ANSWER (programme 75), 1957

MURA-238; Snyder, J. N.
INVARIANT DUCK BUMPS (program 77), 1957

MURA-239
Messy Messy (program 78), 1957

MURA-240; King, R.
FORERUNNER (program 57), 1957

MURA-241; Fosdick, L. D.
Tempermesh (program 76), 1957

MURA-242; Snyder, J. N.
FORMESH field print (program 86), 1957

MURA-243; Snyder, J. N.; Storm, M..R.
Lower binary loader, one card, MU-LBL3, 1957

MURA-244; Snyder, J. N.
Upper binary loader, one card, MU-UBL1, 1957

MURA-245; Snyder, J. N.
Shifting binary loader, one card, MU-SBL2, 1957

MURA-246; Laslett, L. J.
Concerning the y growth phenomenon exhibited by algebraic transformations, 1957

MURA-247; Laslett, L. J.
Supplemental note concerning the algebraic transformations of MURA-246, 1957

MURA-248; Laslett, L. J.; Sessler, A. M.
Stability limit in spiral sector structures near sigma $x = 2\,\mathrm{pi}/4$, 1957

MURA-249; Kerst, D. W.
The relation of eigenpoles to back wound poles, reluctance poles, and other derived pole structures and their winding error effects, 1957

MURA-250; Parzen, G.
The $2\,\mathrm{nu}(y) - \mathrm{nu}(x) = 0$ difference resonance, 1957

MURA-251; Laslett, L. J.
Lapse rate characterizing the convergence of a FOROCYL potential problem, 1957

MURA-252; Laslett, L. J.; Sessler, A. M.
Approximate solutions to the Mathieu equation, 1957

MURA-253; Laslett, L. J.
Optimum spacing of polygonal coils for uniform central field, 1957

MURA-254; Terwilliger, K. M.
A radio frequency system for experiments with the FFAG electron model, 1957

MURA-255; Terwilliger, K. M.; Jones, L. W.
Radio frequency experiments with an FFAG electron model accelerator, 1957

MURA-256; Jones, L. W.
Radial sector accelerators with smaller circumference factor, 1957

MURA-257; Laslett, L. J.
Effect of delta k errors in Illinois model, 1957

MURA-258; Parzen, G.
The equilibrium orbit of a fixed field accelerator, 1957

MURA-259; Van Bladel, J.
Fields in gap excited rectangular ducts, 1957

MURA-260; Jones, L. W.; Pruett, C. H.; Terwilliger, K. M.
Experiments on radio frequency knockout of stacked beams, 1957

MURA-261; Ohkawa, T.
A speculation on a method of propulsion by ion thrust, 1957

MURA-262; Freiser, M. J.
Winding and construction errors in reluctance pole magnets, 1957

MURA-263; Laslett, L. J.
Concerning coupling resonances in the spirally ridged FFAG accelerator, 1957

MURA-264; Ballance, R.; Snyder, J. N.
Binary check sum checker and corrector, one card, MU-CSC2, 1957

MURA-265; Fosdick, L. D.
Octal dump, MU-OCD1, 1956

MURA-266; Zographos, E. M.
Integer dump, MU-IND1, 1956

MURA-267; Snyder, J. N.
Effective address search routine, MU-EAS2, 1956

MURA-268; Snyder, J. N.
Transfer test, visual, MU-TTV1

MURA-269; Snyder, J. N.
Relocatable to absolute translator, MU-RAT1, 1956

MURA-270; Fosdick, L. D.
Fraction dump, MU-FRD1, 1956

MURA-271; Snyder, J. N.
The reflective 704, MU-704R, 1957

MURA-272; Terwilliger, K. M.
Suppression of radio frequency knock-out in stacked beams by phase shifting the betatron oscillations, 1957

MURA-273; Parzen, G.
The linear equations of motion and the tune of a fixed field accelerator, 1957

MURA-274; Snyder, J. N.
Read decimal integer routine, MU-RD11, 1956

MURA-275; Snyder, J. N.
Double precision addition, fixed point, MU-DPA2, 1956

MURA-276; Fosdick, L. D.
Fixed point exponential, BASE-E, MU-EXP1, 1956

MURA-277; Fosdick, L. D.
Fixed point exponential, BASE-2, MU-EXP2, 1956

MURA-278; Zagraphos, E. M.
Binary punch routine, MU-BPU1, 1956

MURA-279; Zagraphos, E. M.
Binary punch routine, MU-BPU2, 1956

MURA-280; Day, E. A.; Mills, F. E.
Vacuum techniques for the range $10^{**}-8$ mm Hg to $10^{**}-10$ mm Hg, 1957

MURA-281
Quarterly report to the Atomic Energy Commission for the period February 6, 1956 to March 31, 1957

MURA-295; Laslett, L. J.
Computational examples of solutions to differential equations which simulate growth of axial oscillations in an FFAG accelerator operated near the $\text{sigma}(x) = 2 \, \text{sigma}(y)$ resonance, 1957

MURA-296; Snyder, J. N.
Formesh (invariants) scope — Programme 113, 1957

MURA-297; Snyder, J. N.
Equicyl scope — Programme 118, 1957

MURA-298; Snyder, J. N.
Invariant Duck Bumps scope — Programme 116, 1957

MURA-299; Snyder, J. N.
TTT scope — Programme 119, 1957

MURA-300; Parzen, G.
Stability limits for the nonlinear resonances in a general spiral sector accelerator, 1957

MURA-301; Van Bladel, J.
Normal modes methods for boundary excited waveguides, 1957

MURA-302; Laslett, L. J.
Round off errors from fixed point linear algebraic transformations computed by IBM 704 program 117, 1957

MURA-309; Fosdick, L. D.
Channel Five (problem 120), 1957

MURA-310; Ohkawa, T.
Electronic devices using plasmas in a magnetic field, 1957

MURA-311; Enoch, J.
Properties of neutralized relativistic electron beams, 1957

MURA-318; Ohkawa, T.
On the two-beam FFAG accelerator, 1957

MURA-319; Mills, R. E.
Resonances of certain Hill equations, 1957

MURA-320; Laslett, L. J.
Computational evidence for a spirally ridged FFAG resonance near 3 sigma(x) + 2 sigma(y) = 2 pi, 1957

MURA-321
Quarterly report to the Atomic Energy Commission for the period April 1, 1957 to June 30, 1957

MURA-328; Snyder, J. N.
FORMESH with STUMBLEBUMPS (program 138), 1957

MURA-329; Stevenson, A. F.
Form of solution of linear differential equations whose coefficients are multiply periodic series, 1957

MURA-330; Terwilliger, K. M.; Jones, L. W.; Pruett, C. H.
Beam stacking experiments in an electron model FFAG accelerator, 1957

MURA-331; King, R.
Variable column integer cathode ray tube display, MU SCP7, 1957

MURA-332; King, R.
Six column fraction cathode ray tube display, MU SCP2, 1957

MURA-333, Storm, M. R.
Fixed point Newton–Cotes quadrature, MU NCI1, 1957

MURA-334; King, R.
Six column integer cathode ray tube display, MU SCP6, 1957

MURA-336; Morton, P. L.
On the design of spiral sector accelerators, 1957

MURA-337-a; Van Bladel, J.
Addendum to report MURA-337: Fields in slanted gap excited rectangular ducts, 1957

MURA-338; Anderson, J. C.
Forsups (programme 99), 1957

MURA-339; Akeley, E. S.
The production by volume current distributions of magnetic fields, which are represented by spherical harmonics in a current free region, which encloses part of the median plane, 1957

MURA-340; Mills, F. E.
Solutions and approximate solutions to a Hill's equation and the Mathieu equation, 1957

MURA-345; Fosdick, L. D.
General alphanumeric cathode ray display, MU SCP3, 1957

MURA-346; McNall, J. F.
Floating point overflow and underflow test, MU OUT7, 1957

MURA-347; Storm, M. R.
Algytee scope (program 121), 1957

MURA-348; Chapman, E. Z.
Cathode ray tube point plotter, MU SCP8, 1957

MURA-360; Anderson, J. C.
A manual for the use of MURA IBM 704 production program

MURA-361; Stump, R.; Waldman, B.
Frequency measurements and resonance survey in the electron model spiral sector FFAG accelerator, 1957

MURA-362; Parzen, G.
The nu = $1/3$ N nonlinear resonance of a fixed field accelerator, 1957

MURA-363; Ling, D. S.
Studies on electron synchroclan, 1956

MURA-364; Jones, L. W.
Notes on the Berkeley Budker Conference, 1957

MURA-365; Parzen, G.
Theory of the nu(y) – nu(x) = 0 nonlinear resonance in alternating gradient accelerators, 1957

MURA-366; Freiser, M. J.
Effects of field bumps due to slotted poles, 1957

MURA-372; Storm, M. R.
Formesh Smallerens (program 159), 1957

MURA-373; Cole, F. T.; Sessler, A. M.
Design of a 50-MeV electron radial sector FFAG accelerator, 1957

MURA-374
Quarterly report to the Atomic Energy Commission for the period July 1, 1957 to September 30, 1957

MURA-375; Cole, F. T.; Day, E. A.; Haxby, R. O.
Parameters and dimensions of the 50-MeV electron FFAG accelerator, 1957

MURA-376; Parzen, G.
The radiation energy loss in a fixed field accelerator, 1957

MURA-377; Cole, F. T.
Field calculations for digital computation with nonscaling dynamics, 1957

MURA-378; Westlund, G. A.
Budbit (programme 157)

MURA-379; Parzen, G.
The nonlinear coupling resonances of an accelerator

MURA-380; Cole, F. T.
Non-Hamiltonian dynamical systems, 1957

MURA-381; Van Bladel, J.
General formulas for gap excited linear ducts, 1957

MURA-382; Storm, M. R.
Coilmate (programme 172), 1958

MURA-383; Bacopoulos, A.; Sessler, A. M.
On a reversed field spiral sector accelerator, 1958

MURA-384; Parzen, G.
Theoretical design of a large two-way radial sector FFAG accelerator, 1958

MURA-385; Jones, L. W.
Space charge limit on beam current in alternating gradient accelerators, 1958

MURA-386; Snyder, J. N.
Gollyconder (programme 171), 1958

MURA-388; Westlund, G. A.
Variable column integer print, MU PRI4, 1957

MURA-393; Parzen, G.
On constructing straight sections in a two-way FFAG accelerator, 1958

MURA-394; Bieri, R. H.
Incertainty in the definition of momentum and emission angle of mesons or hyperons produced in the colliding beam interaction of protons, 1958

MURA-395; Westlund, G. A.
Squave (programme 190), 1958

MURA-396; Jones, L. W.
Some characteristics of the 40-MeV two-way electron model, 1958

MURA-397; Parzen, G.
Theory of accelerators with a general magnetic field, 1958

MURA-401; Storm, M. R.
Scope grid plotter, MU SCO1, 1958

MURA-405; Bieri, R. H.
Lifetime of an electron in a relativistic proton beam, 1958

MURA-406; Cole, F. T.
Scaling radial sector FFAG accelerators without median plane symmetry, 1958

MURA-407; Wallenmeyer, W. A.
Resonance survey in the electron model spiral sector FFAG accelerator, 1958

MURA-408; Smith, J. H.
Report on shielding the MURA high intensity 50-MeV electron accelerator, 1958

MURA-409; Van Bladel, J.
Momentum kicks for particles crossing a gap at an angle, 1958

MURA-410; Bieri, R. H.
Care and feeding of the ultrahigh vacuum system of the two-way model, 1958

MURA-411; Chapman, E. Z.
Ill Tempered Five (program 175), 1958

MURA-412; Jones, L. W.
An FFAG accelerator for achieving 9-BeV center-of-mass collisions between electrons and protons, 1958

MURA-413; Nielsen, C. E.; Sessler, A. M.
Longitudinal space charge effects in fixed field particle accelerators, 1958

MURA-414; Davidson, C. H.; McNall, J. F.
SIFON4 (program 132): simulation of an IBM 650 on an IBM 704, 1958

MURA-415; Kehoe, B.
KEFUNCO (program 206), 1968

MURA-416; Jones, L. W.
Current status of the luminescent chamber, 1958

MURA-417; Westlund, G. A.
Selective audio monitor, MU SAM1 (octal print), 1957

MURA-423; Foster, M.; Guest, G.
A study of the RF phase plane near transition with frequency modulation, 1958

MURA-424; Thorne, G.
General Equations of Motion 1 (GEM1) (program 201)

MURA-425; Symon, K. R.
Report on the high-energy physics conference in Geneva and on a visit to the CERN laboratory, 1958

MURA-426; Johnston, L. H.; Swenson, D. A.; Rowe, E. M.
RF program for the 40-MeV two-way electron model, 1958

MURA-427; Terwilliger, T. M.
Beam extraction from an FFAG synchrotron, 1958

MURA-428; Chapman, E. Z.
Ill Tempered Five: Cyclotron overwrite #1 (program 197), 1958

MURA-430; Jones, L. W.
Comparison of 10 and 15-BeV two-way FFAG accelerator designs, 1958

MURA-431; Parzen, G.
Theory of the AG synchrotron, 1958

MURA-432; Laslett, L. J.
Computational solutions of the two-dimensional wave equation for a cross shaped region, 1958

MURA-433; Mills, R. L.
Liouville's theorem for a continuous medium with conservative interactions, 1958

MURA-434; Morton, P. L.
Effects of radial straight sections on the betatron oscillation frequencies in a spiral sector FFAG accelerator, 1958

MURA-435; Laslett, L. J.
Remarks concerning the interference of higher order Fourier coefficients in the results obtained with the FORANAL program, 1958

MURA-436; Swenson, D. A.
Description of a function generator, 1958

MURA-437; Cole, F. T.
Scaling spiral sector FFAG accelerators without median plane symmetry, 1958

MURA-438; McNall, J. F.
DUCNALL (program 219), 1958

MURA-440; Van Bladel, J.
On Helmholtz's theorem in infinite regions, 1958

MURA-442; Sessler, A. M.
Proposed digital computer program to study an N body problem with one degree of freedom, 1959

MURA-443; Laslett, L. J.
The nonlinear coupling resonance $2 \, \text{nu}(y) - \text{nu}(x) = 1$, 1959

MURA-444; Laslett, L. J.
On the spiral orbit spectrometer, 1959

MURA-445; Laslett, L. J.; Hammer, C. L.
Concerning resonant beam knockout from an A-G synchrotron, 1959

MURA-450; Storm, M. R.
TTT Scope stimulation, 1959

MURA-451; Parzen, G.
Theory of accelerators with a general magnetic field, II, 1959

MURA-452; Laslett, L. J.
Concerning the $\text{nu}/N \rightarrow 1/3$ resonance, I: Application of a variational procedure and of the Moser method to the equation $d^{**}2 \, v/dt^{**}2 + (2 \, \text{nu}/N)^{**}2 \, v + 1/2 \, (\sin 2 \, t) \, v^{**}2 = 0$, 1959

MURA-453; Jones, L. W.
Possible high current 10-BeV proton accelerators, 1959

MURA-454; Parzen, G.
A summary of the linear orbit properties of an accelerator having a general magnetic field, 1959

MURA-455; Symon, K. R.
Report on Brookhaven AGS User's Meeting held in Washington DC, April 29, 1959

MURA-456; Cole, F. T.
Remarks on the design of FFAG accelerators, 1959

MURA-457; Chapman, E. Z.
Change in Ill Tempered Five (program 220), 1959

MURA-458; Becher, A.
Particle–particle collisions in a stacked beam, 1959

MURA-459; Laslett, L. J.
TTITLE = Concerning the nu/$N \rightarrow 1/3$ resonance. 2. Application of a variational procedure and of the Moser method to the equation $d**2\ v/dt**2 + (2\ \text{nu}/N)**2\ v + 1/2\ [\text{sigma}(m=1)\ b(m)\ \sin 2\ mt]\ v**2 + 0$, 1959

MURA-460; Terwilliger, K. M.
Cost estimate comparison of three multi-BeV FFAG accelerators, 1959

MURA-461; Laslett, L. J.
TTITLE = Concerning the nu/$N \rightarrow 1/3$ resonance. 3. Use of the Moser method to estimate the rotation number, as a function of amplitude, for the equation $d**2v/dt**2 + (2\ \text{nu}/N)**2v + 1/2\ (\sin 2t)\ v**2 + 0$, 1959

MURA-462; Cole, F. T.; Haxby, R. O.; Jones, L. W.
High current effects in FFAG accelerators, 1959

MURA-463; Laslett, L. J.
Concerning the nu/N $\rightarrow 1/3$ resonance, IV: The limiting amplitude solution of the equation $d**2u/d\,\text{phi}**2 + (a + B\cos 2\ \text{phi})\ u + B(1)/2\ (\sin 2\ \text{phi})\ u**2 + 0$, 1959

MURA-464; Sessler, A. M.
Energy loss effects and the instability of a coasting beam, 1959

MURA-465; Sands, M.
Ultrahigh-energy synchrotrons, 1959

MURA-466; Van Bladel, J.
Image forces in the third MURA model, 1959

MURA-467; Blewett, M. H.; Courant, E. C.
A high repetition rate 15-BeV Alternating Gradient synchrotron, 1959

MURA-468; Kitagaki, T.
Cost of a 10-BeV scanning field AG synchrotron, 1959

MURA-469; Cool, R. L.
Summary of discussion of the physical interest in higher intensity, 1959

MURA-470; Adair, R. K.; Meyer, D.
Desirable beam characteristics of high intensity accelerators, 1959

MURA-471; Walker, W. D.
Energy considerations for high intensity accelerators, 1959

MURA-472; Good, M. L.; Myer, D.; Richter, B.
Beam separators, 1959

MURA-473; Roberts, A.
Notes on the session on the experimental uses of high intensity accelerators, 1959

MURA-474; Blewett, J. P.
The linear accelerator: A brief summary, 1959

MURA-475; Robinson, K. W.
Multiple frequency accelerators, 1959

MURA-476; Storm, M. R.
DUCK WADDLE (program 249), 1959

MURA-477; Reilly, D.
Computer studies of beam stacking effects, 1959

MURA-478; Mozley, R. F.
The cost of a 10-BeV proton linear accelerator, 1959

MURA-479; Goldwasser, E. L.; Hammer, C. L.
Utilization of very high energies, 1959

MURA-480; Nielsen, C. E.; Sessler, A. M.
Longitudinal space charge effects — Phase boundary equations and potential kernels, 1959

MURA-481; Van Bladel, J.
Sets of eigenvectors for volumes of revolution, 1959

MURA-482; Anderson, J.
Fixed point Runge–Kutta, MU RKY4, 1959

MURA-485; Becher, A.
Interpolation formula with two continuous derivatives, 1959

MURA-486; Hammer, C. L.; Mullin, C. J.; Waldman, B.
Proposed electron–electron scattering experiment, 1959

MURA-487; Terwilliger, K. M.
Achieving higher beam densities by superposing equilibrium orbits, 1959

MURA-488; Nielsen, C. E.; Sessler, A. M.
Longitudinal space charge effects: Influence of energy loss, 1959

MURA-489; Chapman, E. Z.
DRYTEMPER (program 250), 1959

MURA-490; Laslett, L. J.
Concerning the nu/$N \to 1/3$ resonance, IV: A trial function for the limiting amplitude solution of $d**2u/d$ phi**2 $+ (a + b \cos 2$ phi$)\ u + B(1)/2$ (sin 2 phi) $u**2 = 0$, 1959

MURA-491; Lilliequist, C. G.; Symon, K. R.
Deviations from adiabatic behavior during capture of particles into an RF bucket, 1959

MURA-492; Christofilos, N.; Symon, K. R.; Terwilliger, K. M.
Acceleration across the transition energy without increase in amplitude of synchrotron oscillations, 1959

MURA-494; Symon, K. R.
Estimated parameters, performance and cost of colliding beam devices, 1959

MURA-495; Cole, F. T.
Preliminary analytic work on nonscaling spiral sector FFAG accelerators, 1959

MURA-496; Yavin, A. I.
On the advisability of injection from a cyclotron, 1959

MURA-497; Laslett, L. J.; Wolfson, S. J.
Concerning the nu/$N \rightarrow 1/3$ resonance, V: Analysis of the equation $d**2v/ds**2 + (2 \text{ nu}/N)**2v - b/2 (\cos 2 s) v**2 - \text{lambda} (\cos 2 s/3) + 0$, 1959

MURA-498; Parzen, G.
Theory of the fixed frequency cyclotron, 1959

MURA-499; Lichtenberg, D. B.
Background in the neighborhood of a colliding beam region, 1959

MURA-501; Kehoe, B.
SQUAVER II (program 246), 1959

MURA-503; Mullin, C. J.
Beam lifetime in the MURA electron FFAG accelerator, 1959

MURA-505; Storm, M. R.
Roots of a polynomial (program 259), 1959

MURA-506; Lichtenberg, D. B.
Remarks on elementary particles and their interactions, 1959

MURA-507; Smith, W.
Estimate of radiation levels for the MURA 50-MeV electron accelerator, 1959

MURA-508; Westlund, G. A.
Equations of motion SUBROUTINE — EQUAM, 704 FORTRAN program, 1959

MURA-509; Stump, R.
Use of pickup electrode in Wisconsin model, 1959

MURA-510; Stump, R.
The displacement RF system of the Wisconsin model, 1959

MURA-511; Stump, R.
Operating characteristics of the RF system for the Wisconsin model, 1959

MURA-512; Stump, R.
Ferrite cavity model, 1959

MURA-513; Parzen, G.
Beat factors for an accelerator with a general magnetic field, 1959

MURA-514; Dory, R. A.; Morton, P. L.
Effects of radial straight sections on the betatron oscillation frequencies in a spiral sector FFAG accelerator, 1959

MURA-515; Laslett, L. J.
Concerning the $\text{nu}/N \to 1/3$ resonance, VI: Phase plots and fixed points obtained computationally for several equations affected by the $\text{nu}/N \to 1/3$ resonance, 1959

MURA-516; Cole, F. T.; Haxby, R. O.; Symon, K. R.
International conference on high-energy accelerators and instrumentation at CERN, 1959

MURA-517; Haxby, R. O.; Laslett, L. J.; Symon, K. R.
Report on a visit to University College, London, 1959

MURA-518; Cole, F. T.; Haxby, R. O.
Report on a visit to Harwell, England, 1959

MURA-519; Haxby, R. O.; Mills, F. E.; Symon, K. R.
Report on visits to the Center for Nuclear Studies, Saclay, France, 1959

MURA-520; Westlund, G. A.
Printer Paper Plotter, MU PPP5: 704 FORTRAN program, 1959

MURA-521; Westlund, G. A.
Variable column integer or fraction print, MU PIF3: 704 SAP program, 1959

MURA-522; Westlund, G. A.
Variable column floating decimal print, MU PFD3: 704 SAP program, 1959

MURA-523; Westlund, G. A.
Floating point natural logarithm, MU LOG5: 704 FORTRAN II or SAP program, 1959

MURA-524; Westlund, G. A.
Runge–Kutta and print, MU RKP1: 704 FORTRAN II program, 1959

MURA-525; Storm, M. R.
SQUAVER III (program 254), 1959

MURA-526; Westlund, G. A.
SQUAVER IV (program 265), 1959

MURA-527; Chapman, E. Z.
SPIRIT (program 245), 1959

MURA-528; Chapman, E. Z.
GRINDERS (program 264), 1959

MURA-529; Mills, F. E.
Report on a visit to Physikalisches Staatinstitut, Hamburg, 1959

MURA-530; Smith, W.
Energy and angular distribution of the particles resulting from the collision of a 15-BeV proton with a nucleon at rest, 1959

MURA-531; Curtis, C. D.; Rothe, R. E.
Use of an electrolytic tank, 1959

MURA-532; Westlund, G. A.
Floating point Runge–Kutta, MU RKY6: 704 SAP program, 1959

MURA-533; Westlund, G. A.
Universal dump: 704 program, 1959

MURA-534; Westlund, G. A.
HALFDUCK (program F12), 1959

MURA-535; Snowdon, S. C.
Current sources for scaling magnetic fields, 1959

MURA-536; Westlund, G. A.
READ DATA CARDS, MU RCD2: 704 SAP or FORTRAN program, 1959

MURA-537; Jones, L. W.
Magnetogravitational effects in particle accelerators, 1959

MURA-538; Van Bladel, J.
Fields excited by beam currents in circular accelerators, 1959

MURA-539; Snowdon, S. C.
Current sources for scaling magnetic fields (II), 1959

MURA-541; Jones, L. W.
Beam extraction from FFAG synchrotrons

MURA-542; Haxby, R. O.
Experience with a spiral sector FFAG electron accelerator

MURA-543; Staff
The **MURA** two way electron accelerator

MURA-544; Cole, F. T.; Morton, P. L.
Radial straight sections in spiral sector FFAG accelerators

MURA-545; Cole, F. T.
Typical designs of high-energy FFAG accelerators

MURA-546; Laslett, L. J.; Symon, K. R.
Computational results pertaining to use of a time dependent magnetic field perturbation to implement injection or extraction in a FFAG synchrotron

MURA-548; Jones, L. W.; Pruett, C. H.
Comparison of experimental results with the theory of radio frequency acceleration processes in FFAG accelerators

MURA-549; Nielsen, C. E.; Sessler, A. M.; Symon, K. R.
Longitudinal instabilities in intense relativistic beams

MURA-550; Westlund, G. A.
Floating points sin and cosine, MU SIN4: 704 FORTRAN II or SAP program, 1959

MURA-551; Westlund, G. A.
PSEUDONYMER, MU SYM1: 704 program, 1959

MURA-552; McNall, J. F.
CURSES (foiled again), program 283, 1960

MURA-553; Snowdon, S. C.
Magnetic field calculations using distributed currents, 1960

MURA-554; Symon, K. R.
High-energy neutrino experiments with a high intensity FFAG accelerator, 1960

MURA-555; Wong, W. N.
Electromagnetic fields in a donut space, 1960

MURA-556; Wong, W. N.
On Sturrock's perturbation theory, 1960

MURA-557; Storm, M. R.
MU CAD1, complex addition subroutine: Fortran II program, 1960

MURA-558; Bronca, G.
On Robinson's multiple frequency accelerator, 1960

MURA-559; Kerst, D. W.
Electron model of a spiral sector accelerator

MURA-560; Bronca, G.
Adiabatic behavior near transition energy, 1960

MURA-561; Laslett, L. J.; Symon, K. R.
Computational results pertaining to use of a time dependent magnetic field perturbation to implement injection or extraction in a FFAG synchrotron by use of the $nu(r) = N/3$ resonance, 1960

MURA-562; Westlund, G. A.
Sense switch one interrupter — MU SAVE: 704 FORTRAN program, 1960

MURA-563; Rosen, S. P.
Some notes on the kinematics of high-energy nucleon — Nucleon collisions, 1960

MURA-564; Westlund, G. A.
Floating point factorial, MU FACT: 704 FORTRAN II program, 1960

MURA-565; Westlund, G. A.
LOAD BUTTON sequences (MU LBS1, MULBS2, MULBS3): 704 FORTRAN II programs, 1960

MURA-566; McNall, J. F.
CURSES (foiled again) (program 287), 1960

MURA-567; Laslett, L. J.
Schwarz–Christoffel transformations pertaining to magnet edges or peelers, 1960

MURA-568; Christian, R. S.; Snowdon, S. C.
Alternative formulations of magnetostatic problems, 1960

MURA-569; Akeley, E. S.
The production by surface currents of magnetic fields suitable for spiral ridge accelerators, 1960

MURA-571; Westlund, G. A.
XMASH (program F16), 1960

MURA-573; Westlund, G. A.
Floating point exponential, MU EXP3: 704 FORTRAN program, 1960

MURA-574; Rosen, S. P.
On the neutrinos emitted in beta decay and mu capture, 1960

MURA-575; Bronca, G.
Acceleration across transition energy, 1960

MURA-576; Richter, B.
Proposed radio frequency separator, 1959

MURA-578; Dory, R. A.
Constant area RF buckets in the IBM 704 programme TTT, 1960

MURA-580; Westlund, G. A.
Positive random integer generator, 1960

MURA-581; Westlund, G. A.
MISHMASH (program F23) and DATA CHECK (program F28), 1960

MURA-582; Westlund, G. A.
FRANCIS (program F27), 1960

MURA-583; Snowdon, S. C.
Integral scaling magnetic field using distributed windings, 1960

MURA-584; Good, M. L.
Electromagnetic separation of secondary beams at the MURA accelerator, 1960

MURA-586; Cristal, B.
Modified Bessel functions of the first kind, order zero, 1960

MURA-587; Storm, M. R.
FOURIER-SUM (program F43), 1960

MURA-588; Dickman, D.
COSPLASH (program 308), 1960

MURA-589; Gordon, M. M.
Linear oscillations about off center, fixed point orbits, 1960

MURA-590; Jones, L. W.
A neutrino beam design, 1960

MURA-591; Snowdon, S. C.
Relaxation calculation of integral scaling magnetic fields produced by distributed currents, 1960

MURA-593; Blosser, H. G.; Gordon, M. M.
Performance estimates for injector cyclotrons, 1960

MURA-594; Denman, H. H.
Convergence rate of the nine point extrapolated Liebmann algorithm, 1960

MURA-595; Lassettre, C. A.
Computational studies of coupling resonances in spirally ridged accelerators, 1961

MURA-596; Sessler, A. M.
Theoretical remarks concerning the 3 sigma(x) + 2 sigma(y) = 2 pi resonance in spiral sector accelerators, 1961

MURA-597; Dickman, D.
One13 (program 312), 1961

MURA-602; Snowdon, S. C.; Christian, R. S.
Relaxation calculation of magnetic fields produced by distributed currents in presence of iron with a variable permeability, 1961

MURA-603; Carlson, H. L.
FORTRAN Fourier analysis routine, MU FAN1: 704 FORTRAN program, 1960

MURA-604; Chapman, E. Z.
FLEXIBLE FIVER (program 280), 1960

MURA-606; Boilen, J. B.
The effect of nearby buckets on bucket area in RF acceleration, 1961

MURA-611; Wong, W. N.
Time modes of a coasting beam in a simple perturbation approach, 1961

MURA-612; Wong, W. N.
Effects of radial straight sections from a general point of view, 1961

MURA-617; Westlund, G. A.
UNICYL (program F47), 1961

MURA-618; Peekna, A.
Experimental stress analysis of a circular bubble chamber window with a beveled mounting

MURA-619; Van Bladel, J.
Screening with current sheets

MURA-620; Bjerke, C. C.
Sort by column and character, MU SRTC: 1401 program

MURA-621; Carlson, H. L.
SHARE abstract listing routine, MU X1: 1401 program

MURA-622; Edwards, T. W.
Proton linear accelerator cavity calculations

MURA-623; Van Bladel, J.
The effect of curvature on the fields in a circular accelerator

MURA-624; Storm, M. R.
BCD cards to tape program, MU CTH1: 1401 program

MURA-625; Storm, M. R.
Sequential numeric card sort, MU SRT2: 1401 program

MURA-626; Bjerke, C. C.
Row binary card dump, MU RBCD: 1401 program, 1961

MURA-628; Dory, R. A.
80 column card lister, X10: 1401 SPS program, 1961

MURA-629; Anderson, J. C.; Parzen, G.
The Parmesh dynamics program, 1961

MURA-630; Anderson, J. C.; Parzen, G.
The Parmesh dynamics program, 1961

MURA-631; Dory, R. A.
704 SAP program, 1961

MURA-633; Van Bladel, J.
Some remarks on force free coils, 1961

MURA-634; Parzen, G.
Perturbation theory for accelerators with a general magnetic field, 1962

MURA-635; Anderson, J. C.; Parzen, G.
The PARMESH dynamics program III, 1962

MURA-636; Bush, G. E.
COILFINDER, 1962

MURA-637; Van Bladel, J.
The excitation of a resonant cavity by waveguides of small cross-sectional dimensions, 1962

MURA-642; Edwards, T. W.
MESSYMESH (program F46), 1962

MURA-644; Bjerke, C.
The excitation of a circular cylindrical cavity by a circular waveguide, 1962

MURA-645; Swenson, D. A.
Computer program for beam transport problems, 1962

MURA-649; Morin, D. C., Jr.
Transverse space charge effects in particle accelerators, 1962

MURA-660; Edwards, T. W.
MEMORY DUMP/SAVE subroutines — MU SAVE: 704 FORTRAN subroutines coded in FAP, 1962

MURA-661; Wong, W. N.
On the stability of orbits in a spiral magnetic field, 1962

MURA-667; Parzen, G.
Perturbation theory for accelerators with a general magnetic field. 2, 1963

MURA-669; Edwards, T. W.
Explicit solution of the general cubic equation — MU CUBIC: 704 FORTRAN subroutine, 1963

MURA-673; Parzen, G.; Morton P. L.
Effects of field perturbations in FFAG accelerators, 1963

MURA-674; Scott, D.
A method of radio frequency inflection into a particle accelerator, 1963

MURA-679; Morton, P. L.
Particle dynamics in linear accelerators, 1963

MURA-698; Galonsky, A.
Neutrino fluxes without focusing and with "ideal" focusing, 1964

MURA-700; Snowdon, S. C.
Study of a 500-MeV high intensity injector, 1964

MURA-701; Rowe, E. M.
On the suitability of the 500-MeV FFAG synchrotron as an injector for the AGS and LGS, 1964

MURA-702; Edwards, T. W.
Divide by zero locater — MU DIVZR: 704 SAP subroutine, 1964

MURA-704; Meads, P. F., Jr.
Floating point Runge–Kutta, MU RKY7, 1964

MURA-706; Snowdon, S. C.
Algorithms for relaxation calculation of FFAG magnetostatic fields, 1965

MURA-707; Curtis, C. D.; Lee, G. M.
Preaccelerator column design, 1965

MURA-708; Kreiegler, F. J.; Snowdon, S. C.
Magnetic field calculations suitable for a heavy liquid bubble chamber magnet, 1965

MURA-709; del Castillo, G.
A low field superconducting FFAG model magnet, 1965

MURA-711; Lee, G. M.; O'Meara, J. E.; Winter, W. R.
Preliminary design development of a vacuum chamber for the Argonne ZGS, 1965

MURA-712; Meads, P. F., Jr.
The numerical design of a resonant extraction system for the MURA 50-MeV electron accelerator, 1965

MURA-713; Austin, B.; Edwards, T. W.; O'Meara, J. E.; Palmer, M. L.; Swenson, D. A.; Young, D. E.
The design of proton linear accelerators for energies up to 200 MeV, 1965

MURA-714; Mills, F. E.; Curtis, C. D.; Swenson, D. A.; Young, D. E.
Proceedings of the Conference on Proton Linear Accelerators Madison, Wisconsin, 1964

MURA-716; Bush, G. E.
High speed digital dynamic field measuring system, 1965

MURA-717; Dory, R. A.
Write up of MURA 704 program F51: Particle space charge program, 1965

MURA-718; Dayton, B.; Mills, F. E.; Radmer, C.; Jones, L. W.; Camerini, U.; Good, M. L.; Subramanian, A.
A search for massive particles in cosmic rays, 1966

MURA-719; Chien, C. S.
Analytical calculation of FFAG operating points and orbits, 1966

MURA-722; Baumann, C. A.; Camerini, U.; Fry, W. F.; Gams, M. C.; Hilden, R. A.; Laufenberg, J. F.; Palmer, M. L.; Powell, W. M.; Sviatoslavsky. I. N.; Winter. W. F.
Bubble chamber research with the MURA model heavy liquid chamber, 1967

MURA-723; Owen, C. W.; Radmer, C. A.; Young, D. E.
RF perturbation measurements in long linac cavities, 1966

MURA-724; Buer, H. H.
Address labeling program — MU LABL: 1401 SPS program X71, 1966

MURA-725; Lyon, D.E., Jr.; Subramanian, A.
Design of ionization spectrometers using iron scintillators for the detection of hadrons in the 100–1000 BeV range, 1967

APPENDIX C

MURA ARCHIVES

Archives containing MURA material can be found at the following places [Kinraide, 2000]:

University of Minnesota Archives
University of Wisconsin–Madison Physical Sciences Laboratory
Fermi National Accelerator Laboratory
Michigan State University
Northwestern University
Purdue University
Ohio State University
University of Iowa
University of Michigan
University of Illinois at Urbana-Champaign
University of Notre Dame

Descriptions of MURA materials in these archives follow. Much of this information is from the files of Rebecca Kinraide [Kinraide, 2000]. It will be noted that the university archives generally contain the files of their presidents during the MURA period, which contain correspondence and other materials, particularly concerning MURA business matters. Very few contain files of faculty members, which would be a better source of technical materials. An almost complete collection of MURA materials, particularly of official files, is contained in the University of Minnesota Archives. The Physical Sciences Laboratory at the University of Wisconsin has a fairly complete set of technical materials.

- University of Minnesota Archives
  218 Elmer L. Anderson Library, University of Minnesota, 222 21st Ave. S., Minneapolis, MN 55455, USA
  Midwestern Universities Research Association Papers, 1953–1974.

The Minnesota archives contain the most extensive collection of MURA papers. By action of the MURA Board of Directors and the individual university members, the University of Minnesota Archives was designated the depository for all MURA corporate records, with the exception of contract, fiscal, and audit records which are deposited at the Physical Sciences Laboratory, University of Wisconsin–Madison.

The MURA Archive at Minnesota contains, in the following order:

(I)   Legal Documents

      (A)   Organizational Papers
      (B)   Patents
      (C)   Internal Revenues
      (D)   Contracts and Deeds

            (1)   Organizations
            (2)   Employment

(II)   Organizational Committee of Midwestern Universities Research Association, Inc.
(III) Board of Directors

      (A)   Lists of Boards of Directors by Fiscal Year, 1955–1967
      (B)   Annual Meetings
      (C)   Other Meetings
      (D)   Meetings Involving Other Organizations

(IV) Individual Members

      (A)   Lists of Individual Members by Fiscal Year, 1957–1972
      (B)   Annual Meetings
      (C)   Special Meetings

(V)   Executive Committee
(VI) Associated Midwest Universities
(VII) Correspondence and General Files 1953–1974
(VII) Site Data

      (A)   Site Proposals Submitted to MURA, 1956
      (B)   Correspondence re Site Selection, 1955–1958
      (C)   Site Reports

(IX) Reports and Proposals
(X)   Related Materials (including some materials from other laboratories)

A detailed list of MURA items in the Minnesota archives can be found in materials collected by Rebecca Kinraide located at the Physical Sciences Laboratory of the University of Wisconsin–Madison (see below).

- University of Wisconsin–Madison Physical Sciences Laboratory
  3725 Schneider Dr. Stoughton, WI 53589–3098, USA

  In addition to the MURA corporate contract, fiscal, and audit records, the Physical Sciences Laboratory contains photographs collected by the MURA working group, miscellaneous materials (minutes, correspondence, reports from external sources, etc.), letters and reports from MURA and the Atomic Energy Commission, MURA proposals, computer programs, technical tables, miscellaneous technical reports, newsletters and conference reports, clippings, and MURA reports and reprints, plus employee files, archives submitted by a few MURA personnel and materials collected by Rebecca Kinraide, a student archivist [Kinraide, 2000]. Kinraide's materials include oral interview tapes, videotapes, archivist letters from MURA universities, responses to questionnaires, and other materials.

  The University Division of Archives, in its division of College of Letters and Science, subdivision Department of Physics, has personal files from 1953 to 1967 (five boxes) submitted in 1979 by Keith Symon, as well as files submitted by Donald Kerst.

- Fermi National Accelerator Laboratory
  Archivist, History of Accelerators Project
  Wilson Hall 3-SE, MS-109, PO Box 500, Batavia, IL 60510, USA
  Fermilab has an almost complete collection of MURA Reports.

- Michigan State University
  Archives and Historical Collections
  101 Conrad Hall, East Lansing, MI 48824-1327, USA
  The archivist was unable to locate any MURA materials.

- Northwestern University
  University Archives
  Northwestern University Library
  1935 Sheridan Road, Evanston, IL 60208-2300, USA

  A file with the title "Midwestern Universities Research Association — Argonne Universities Association" appears in the records of Payson S. Wild, former Vice President and Dean of Faculties (Box 13, Folder 1). It contains about a half inch of correspondence and related items which date between 1954 and 1971.

- Purdue University
  The Libraries Special Collections Department
  1530 Stewart Center, West Lafayette, IN 47907-1530, USA

  The Special Collections Department contains no materials related to MURA. It does have the papers of Frederick Hovde, President of the University from 1946 to 1971, which may contain some MURA correspondence or other materials.

- Ohio State University
  University Archives
  2700 Kenny Road, Columbus, OH 43210, USA

  The archives contain the papers of Novice Gail Fawcett, President of the University from 1956 to 1972, and of his predecessor, Howard L. Bevis, whose papers also have a folder on MURA. In additions, there is a "Centennial History of the Physics Department" which describes OSU's involvement in MURA from 1955 until 1969 (when the history was written).

- University of Iowa
  The University Libraries
  100 Main Library, Iowa City, IA 52242-1420, USA

  They have some folders on Francis Cole and Joseph Jauch which contain correspondence and other papers relating to MURA. They also have the office files of Virgil Hancher, President of the University of Iowa from 1940 to 1964, containing considerable material on MURA.

- University of Michigan
  Bentley Historical Library
  1150 Beal Avenue, Ann Arbor, MI 48109-2113, USA

  A substantial collection of official MURA papers (minutes, proposals, organization, correspondence, committee reports, etc.) is among the papers of Harlan Hatcher, President of the university during the MURA period. There are eight boxes in the Kent Terwilliger collection, including photographs and technical reports. There is a small collection of Department of Physics papers which contains some mention of H. Richard Crane and Lawrence Jones, who were active in MURA.

- University of Illinois at Urbana-Champaign
  University Library, University Archives
  19 Main Library, 1408 West Gregory Drive, Urbana, IL 61801, USA

  MURA material is included in the papers of Donald Kerst and P. Gerald Kruger. There are probably also references to MURA in the administrative records of the President and Provost.

- University of Notre Dame
  University Archives
  607 Hesburgh Library, University of Notre Dame, Notre Dame, IN 46556, USA

The archives have two file drawers of MURA materials in the files of Father Theodore Hesburgh, President. They contain Articles of Incorporation and By-Laws, contracts, minutes, annual and quarterly reports, site selection materials, proposals, correspondence, and Associated Midwest Universities materials. There are a few MURA papers, primarily business materials, in the files of Bernard Waldman.

Appendix D

MURA PERSONNEL

Personnel are listed according to their affiliation at the time they were associated with MURA. "Seniors" refers to the well-known senior physicists at MURA universities who played a role in the organization and direction of MURA. "Scientific Staff" refers to professional physicists who were on the MURA payroll. "Engineers" refers to technical support staff members, also on the MURA payroll. "Administrative Staff" refers to administrators and office people, including secretaries, who played an important role in the MURA program. "Technicians" refers to people who played various technical roles. "Students" refers (primarily) to graduate students who worked on the MURA projects. "Participants" refers to physicists who were not on the MURA payroll, but who made contributions to MURA through their participation as members of workshops, summer studies, at meeting discussions, etc. "Visitors" includes visitors who should be noted, although they may not have contributed directly to the MURA program. Not included in these lists are many visitors to MURA, attendees at various meetings, and visitors to MURA laboratories.

The authors have tried to acknowledge persons who have contributed to MURA history but undoubtedly they have overlooked some in this listing. They regret these omissions.

After some names, the book section in which that individual is first mentioned is given in parentheses.

SENIORS

Donald W. Kerst: University of Illinois (1)
P. Gerald Kruger: University of Illinois (2.3)
H. Richard Crane: University of Michigan (2.1)
John Williams: University of Minnesota (3.1)
Ragnar Rollefson: University of Wisconsin (3.1)
Allan Mitchell: University of Indiana (3.1)
L. Jackson Laslett: Iowa State University (3.1)

Bernard Waldman: University of Notre Dame (4.9)
Josef Jauch: University of Iowa (4.1)
Robert O. Haxby: Purdue University (3.1)
Lawrence R. Lunden: University of Minnesota (4.6)
A.W. Peterson: University of Wisconsin (4.6)

SCIENTIFIC STAFF

Anderson, Jess C.
Austin, Bonnie
Bieri, Rudy E. H.
Bush, George E.
Carlson, Henry
Chang, Tsu-Shen
Christian, Richard (4.4)
Cole, Frank (3.1)
Curtis, Cyril (5.1)
Dickman, Donald
Del Castillo, Gustavo
Enoch, Jacob
Fasolo, J.
Fast, Ronald
Fosdick, Lloyd
Freiser, Marvin J.
Galonsky, Aaron
Henkel, Jane
Hicks, John
Jones, Lawrence (3.1)
Kehoe, Brandt
Kriegler, Frank J.
Marty, Paul R.
McGruer, James
McNall, John F.
Meads, Philip F., Jr.
Mills, Fred (3.5)
Morin, Jr. Dornis C. (Bud)
Mulady, James R.
Nielsen, Carl (3.10)
Ohkawa, Tihiro: Japan (3.2)

Owen, Curtis (5.1)
Otte, Roger
Parzen, George
Peekna, Andre
Pruett, Charles (3.4)
Ralph, William W.
Rosen, S. Peter
Rowe, Ednor (3.4)
Sessler, Andrew (3.3)
Shea, Michael F. (3.5)
Snowdon, Stanley C. (3.8)
Snyder, James N. (3.1)
Steben, John
Storm, Melvin R.
Stump, Robert F.
Swenson, Donald A. (5.1)
Symon, Keith R. (3.1)
Terwilliger, Kent M. (3.1)
Vogt Nilsen, Nils (3.3)
Wallenmeyer, William (3.5)
Wong, Wen Nong
Yavin, Avivi I.
Young, Donald (3.11)

ENGINEERS

Beck, Robert
Boatner, Henry
Baumann, Carl A.
Berndt, Martin
Cwayna, James R.
Day, Edward
Hagen, James
Hicks, John
Hilden, Richard
Hollar, Ed L.
Jergens, Richard
Kennedy, Ed
Laukart, Erich

Lee, Glenn
Mandley, Irving J.
O'Meara, John
Palmer, Max
Peterson, Frank, (3.5)
Radmer, Carl (3.5)
Shultz, Otfried P.
Sviatoslavsky, Igor N.
Van Bladel, Jean
Winter, William (4.9)

ADMINISTRATIVE STAFF

Keith, Marshall (4.9)
Wittig, Harold
Anderson, Arloa
Gehring, Rose
Gravine, Bob
Husker, Esther
Kotzke, Fran
Latzke, Dave
Luniak, Jean
Mandley, Eileen
Mulrooney, Rosemary
Olson, Esther
Olson, Nancy
Onken, Nancy
Pazorski, Nancy
Reilly, Phil
Shalton, Paul
Sweeny, Jean
Tortorici, Carmen

TECHNICIANS

Brown, Roswell
Butler, William
Fasking, Richard

Ford, Doyle
Henderson, Glen
Hogan, James
Hurst, Edmond
Laufenberg, John F.
Lawry, Erwin
Posey, Steven
Severson, Weston
Silverling, Leon
Springstube, August
Wildenradt, Jan
Wille, Edwin

STUDENTS

Binford, Thomas: University of Wisconsin
Dory, Robert A.: University of Wisconsin
Edwards, Terry
Foster, Margaret: University of Wisconsin
Kriegler, Frank: University of Wisconsin
Lilliequist, Carl G.: University of Wisconsin
Meier, Homer K.: University of Wisconsin
Mogford, James: University of Wisconsin
Morton, Philip: Ohio State University
Owen, Curtis: University of Wisconsin (3.4)
Roiseland, Don S.: University of Wisconsin (3.6)
Trzeciak, Walter
Wilkinson, David: University of Michigan (3.4)
Chapman, Elizabeth Zographos

PARTICIPANTS

Adair, Robert K.: Yale University and BNL
Adler, Felix: Carnegie Tech (3.3)
Akeley, Edward: Purdue University (3.3)
Blewett, John P.: BNL (2.3)
Blewett, M. Hildred: BNL (3.11)
Blosser, Henry G.: Michigan State University (6.1)

Bronca, Gaston
Bruck, Henri: Saclay (France)
Christofilos, Nicholas: Livermore National Laboratory (2.3)
Cool, Rodney L.: BNL
Courant, Ernest: BNL (2.3)
Dalitz, Richard H.: University of Chicago
Doty, R.: Iowa State University
Francis, Norman: University of Indiana (3.1)
Frisch, Otto: England (3.3)
Fulbright, W.: University of Rochester
Goldwasser, Edwin L.: University of Illinois
Good, Myron L.: University of Wisconsin (5.6)
Hammer, Charles L.: Iowa State University
Hansen, Albert O.: University of Illinois
Hildebrand, Roger H.: ANL
Horwitz, Norman: LBNL
Johnston, Lawrence: University of Minnesota (3.1)
Judd, David: LBNL (3.3)
King, R. F.: University of Illinois
Kitagaki, Toshio: Princeton-Penn Accelerator (3.9)
Knipp, J. K.: Iowa State University
Lax, P.: New York
Lichtenberg, Donald B.: University of Indiana (3.9)
Meyer, D.: University of Michigan
Mozley, R. F.: SLAC
Mullin, C. J.: University of Notre Dame
O'Neill, Gerard K.: Princeton University (3.9)
Palfrey, Thomas R.: Purdue University
Post, Richard F.: Livermore National Laboratory
Powell, John: University of Wisconsin (3.1)
Richter, Burton: SLAC
Roberts, Arthur: University of Rochester
Robinson, Kenneth W.: CEA
Rohrlich, Fritz: University of Iowa (3.1)
Sands, Matthew: Cal Tech (4.6)
Scalise, Ted: LBNL
Schein, Marcel: University of Chicago
Schoch, Arnold: CERN
Shoemaker, Frank C.: Princeton University
Tautfest, George: Purdue University

Thom, R.: Sao Paolo, Brazil
Tollestrup, Alvin V.: Cal Tech
Walker, William D.: University of Wisconsin (5.6)
Wattenberg, Al: University of Illinois
Welton, Ted A.: Oak Ridge National Laboratory
Wright, S. Courtenay: University of Chicago (3.1)
Ypsilantis, Tom: LBNL
Zaffarano, Daniel: Iowa State University (3.1)

VISITORS

Brueckner, Keith A. (3.1)
Budker, Gersch I.: INP, Novosibersk (USSR) (3.11)
Camerini, Ugo: University of Wisconsin (5.6)
Crosbie, Edward: ANL (3.3)
Firentz, Melvin: ANL (3.3)
Foss, Martin: ANL
Fry, William F.: University of Wisconsin (5.6)
Hamermesh, Morton: ANL (3.1)
Hofstadter, Robert: Stanford University (3.1)
Jauch, Joseph: University of Iowa (3.1)
Kolomensky, Andrei: Lebedev Inst., Moscow, Russia (3.2)
Livingood, John: ANL (3.3)
Livingston, Stan: BNL (2.3)
McMillan, Edwin: LBNL (2.1)
Newton, Roger G.: University of Indiana (3.9)
Panofsky, Wolfgang: SLAC (3.1)
Prome, M.: Saclay (France)
Ross, Marc H.: University of Indiana (3.9)
Sachs, Robert G.: University of Wisconsin (3.1)
Smith, J. H.: University of Illinois
Snyder, Hartland: BNL (2.3)
Taub, Abraham: University of Illinois (3.1)
Taylor, Clyde: CERN
Teng, Lee: ANL (3.3)
Von Tersch, W.: Iowa State University.
Walkinshaw, William: Harwell RHEL, England
Watson, Kenneth M.: University of Wisconsin (3.1)

LIST OF ILLUSTRATIONS WITH ACKNOWLEDGMENTS

Illustrations without an explicit credit line are generally from the MURA archives or the personal collections of the authors.

Cover: *New York, New York.*
Courtesy of Sandi Adams (www.sandiadams.com)

Fig. 1.1: The Radial Sector Model at Michigan, 1956.
Courtesy of the University of Michigan photo archives.

Fig. 1.2: The Spiral Sector Model at Madison, 1958.
Courtesy of the MURA photo archives.

Fig. 1.3: The 50 MeV Two-Way Model at the Stoughton Site, 1961.
Courtesy of the MURA photo archives.

Fig. 3.1: The 1953 MAC Summer Study group in Madison.
Courtesy of the University of Wisconsin.

Fig. 3.2: A close-up photograph of the Radial Sector Model with one magnet removed.
Courtesy of the University of Michigan photo archives.

Fig. 3.3: A radial-focusing magnet from the Radial Sector Model.
Courtesy of the University of Michigan photo archives.

Fig. 3.4: The "necktie" stability diagram for a radial sector FFAG accelerator.

Fig. 3.5: The Michigan Working Group, autumn 1954.
Courtesy of the University of Michigan photo archives.

Fig. 3.6: Members of the 1955 MURA Summer Study in Ann Arbor.
Courtesy of the University of Michigan photo archives.

LIST OF SIDEBAR PORTRAITS

Name	Acknowledgment
Donald William Kerst	Courtesy of University of Illinois
Tihiro Ohkawa	Courtesy of Tihiro Ohkawa
Lawrence Jackson Laslett	Courtesy of Lawrence Berkeley National Laboratory
Francis T. Cole	Courtesy of Fermilab
Charles Pruett	Courtesy of Charles H. Pruett
Kent M. Terwilliger	Courtesy of Jens Zorn, Department of Physics, University of Michigan
Robert O. Haxby	Courtesy of Physics Department, Purdue University
Carl Radmer	Courtesy of Ethel Radmer
Richard Christian	Courtesy of Physics Department, Purdue University
Stanley C. Snowdon	Courtesy of Fermilab
James N. Snyder	Courtesy of Physics Department, University of Illinois
Carl E. Nielsen	Courtesy of Ohio State University
H. Richard Crane	Courtesy of Jens Zorn, Department of Physics, University of Michigan
Marshall W. Keith	Courtesy of F. E. Mills
William Anton Wallenmeyer	Courtesy of W. A. Wallenmeyer
William R. Winter	Courtesy of Physical Sciences Laboratory, UW–Madison
Bernard Waldman	Courtesy of M. F. Shea
Cyril D. Curtis	Courtesy of Charles Schmidt
Donald A. Swenson	Courtesy of D. A. Swenson
Ednor M. Rowe	Courtesy of Physical Sciences Laboratory, UW–Madison
Henry G. Blosser	Henry G. Blosser
Fishing Trips	F. E. Mills

APPENDIX F

THE RAMSEY PANEL, LYNDON JOHNSON, AND THE END OF MURA, AS SEEN IN WASHINGTON

David Z. Robinson

In March 1961, I joined the White House Staff to work for Jerome Wiesner, who was President Kennedy's Science Adviser, and Chair of the President's Science Advisory Committee (PSAC). I was on a leave from Baird-Atomic, Inc., the company I had joined in 1949 after I obtained my Ph.D. in Chemical Physics from Harvard.

There were about fifteen members of PSAC, most of whom served three-year terms. PSAC met as a group at least two days a month, with ad hoc Panels working hard in between meetings.

THE RAMSEY PANEL

In the summer of 1962, our office was faced with three accelerator proposals, one for construction of a high intensity 10 Bev FFAG accelerator from MURA, one for advanced design of a 100 BeV accelerator from the Lawrence Berkeley Laboratory (LBL), and one for a large preliminary design study for a 600–1000 Bev accelerator from Brookhaven. The Bureau of the Budget (BOB) was concerned and unenthusiastic about all three projects. Accordingly, in November 1962, PSAC agreed to another joint PSAC-GAC Joint Panel, chaired by Norman Ramsey of Harvard. (See Chapter 4 for a complete description of the members and their recommendations.) Previous PSAC-GAC Panels had been influential in the decision to build the accelerator at the Stanford Linear Accelerator Center (SLAC). Randall Robertson of the NSF and I were *Ex Officio* members. I attended and participated in all the meetings of the Panel. I also drafted two sections of its Report but did not vote on any of its decisions. Almost all the meetings were held at PSAC Offices in the Executive Office Building (EOB) next door to the White House.

Wiesner addressed the Panel at its first meeting. He described the difficulties in getting final approval of the SLAC accelerator, citing problems both with the BOB and with Congress. The total Federal Budget was under $100 billion. The deficit was described publicly to the nearest hundred million dollars. (For example, it would be announced as $3.2 billion.) Adding a new accelerator costing over $100 million (the cost of the SLAC accelerator) would increase the announced deficit, and was certain to get the President's scrutiny. Wiesner said that it was highly unlikely that we could get authorization of a new accelerator until at least two years had passed. An additional expensive accelerator would require at least two more years.

The Panel had fourteen meetings in six months. It reviewed all proposed activities in high-energy physics. Its members included high-energy experimenters (Chamberlain, Goldwasser, Panofsky, and Williams), high-energy theorists (Gell-Mann and Lee), and physicists with other specialities (Ramsey, Abelson, Purcell, and Seitz). Phil Abelson — head of the Carnegie Institution of Washington, and the Editor of *Science* — was a known skeptic about the value of high-energy physics research for society as a whole. The geographic and scientific distribution of the members — and above all their prestige as scientists — gave particular weight to their conclusions.

I recall Ed Purcell's strong and persuasive view that new costly accelerators required a substantial expansion of the capabilities of present ones. He felt strongly that the proposed energy of 100 BeV in the original LBL design was too low, and pushed to have us recommend going to at least 200 BeV. He was also eloquent on the potential importance of colliding beams.

To me, Murray Gell-Mann was the most persuasive supporter of the MURA project as a priority. He was a theoretical physicist, and thus did not have a special interest in the outcome, and he was not affiliated with any of the organizations making the proposals.

In the end, the Panel felt that a move to substantially higher energy had greater priority than a move to substantially higher intensity. However the MURA high intensity proposal was much further along than the LBL high-energy proposal. The Panel felt that both accelerators were worth building, but it was quite aware of Wiesner's comments about the expected time between authorizations of large new accelerators. After much deliberation, the Panel report of May 10, 1963 (its full recommendations are given in Chapter 4) recommended authorization, *at the earliest possible date*, of the construction by LBL of a 200 BeV accelerator. They also recommended authorization in the next fiscal year (1965) of the MURA accelerator *without permitting this to delay the steps toward higher energy*. The report was unanimous, although Abelson changed his position in an editorial a couple of years later. The members of the Panel did recognize that by setting priorities the way they did, they might doom the MURA accelerator.

When the Report came before PSAC, it was generally accepted. I. I. Rabi, however, pointed out that almost all panel members were not active accelerator users. He persuaded PSAC to appoint a group of younger users, giving as examples his Columbia colleagues, Leon Lederman and Mel Schwartz. This second Panel was chaired by Myron "Bud" Good of Brookhaven. I attended their meetings, and in addition to endorsing the Ramsey Panel Report, they spent much of their discussions on the need for access, local support, and even a voice in management for outside users.

THE JOHNSON DECISION

Budget decisions for the fiscal year 1965 (which went from July 1, 1964 until June 30, 1965) were received from the Departments and Agencies in August 1963. The Agency budgets were reviewed, first by the junior BOB staff members, and finally in a senior staff meeting chaired usually by Deputy Director Elmer Staats or Director Kermit Gordon. OST staff sat in on these reviews when technical issues were involved. (It was the only part of the Executive Office of the President with that privilege.) The AEC submitted the request for authorization as part of its budget. I sat in on the MURA discussions, reiterating Wiesner's support of the Ramsey Panel recommendations, which included both the support of MURA and the conditions placed by the Panel. These particular discussions started in November 1963.

The afternoon of Friday November 22, 1963, while we were mourning President Kennedy's assassination, Wiesner received a call from Kermit Gordon. Gordon told him that he had talked to President Johnson who was on his way back to Washington from Texas, and that Johnson wanted him to continue at full speed to work on the Budget for FY65, his first major responsibility as the new President. Thus the Budget discussion with Wiesner and me that had been scheduled for Saturday morning was still on.

I came into the EOB on Saturday morning at about 9:30 AM and walked by Johnson's Office when he had been Vice-President. The anteroom had already been stripped of its old furniture, and there was a pile of other furniture in the hall, which included Evelyn Lincoln's desk. (She was President Kennedy's personal secretary.) "*Le roi est mort. Vive le roi!*" went through my mind. We worked with BOB people all day Saturday and Sunday on a number of matters to get material ready for the new President. At that point, as before, Gordon was dubious about the MURA proposal. I do not believe that Kennedy had had a real opportunity to review the AEC budget or to take a final position on MURA.

Wiesner had decided in August of 1963 to go back to MIT as Dean of Science, beginning at the end of January. Kennedy had chosen (at Wiesner's suggestion) Donald Hornig, a PSAC member, whom Kennedy did not know, although they were

Harvard classmates. After the assassination, Wiesner informed Johnson of his plans, but offered to stay on for a few months, if Johnson wished it. OST had worked reasonably effectively with Johnson's people, particularly on issues involving the Space Council, which Johnson chaired. However, Johnson and Wiesner were not close the way that Kennedy and Wiesner were. For instance, when Kennedy would go to Cape Cod for summer weekends, Wiesner would often hitchhike on Air Force One to get to his summer home in Martha's Vineyard. President Johnson told Wiesner that he would follow Kennedy's plan, and take Hornig as his Science Adviser in February.

The MURA University Presidents had mounted a major campaign with their Senators and Congress Members starting in the summer. These efforts were increased in December, and Johnson received many communications promoting MURA. At the same time, Johnson was insisting to Gordon that the budget be kept below $100 billion and it was drifting over it. In early December, Gordon formally recommended that MURA not be funded, listing as a major reason that the 200 BeV accelerator was higher priority, and that the authorization of the MURA accelerator would delay it. Meanwhile, Senators from the MURA states, particularly Dirksen, Humphrey, and Douglas were putting pressure on President Johnson. Johnson scheduled a meeting for December 20, which would include individuals from MURA, Senators from MURA States, Wiesner, Gordon, and others involved from the Executive Office.

Two days before the meeting, Johnson asked Wiesner to prepare two one-page memos, one giving ten reasons to support the MURA construction, and one giving ten reasons against support. I helped him with the preparation. They were both signed by him, and on his stationery.

I did not attend the meeting, but I saw Wiesner immediately afterward. He was shaken. For many years, he had been an adviser to Hubert Humphrey, particularly on arms control issues, and the Wiesners and Humphreys had spent evenings together. I am sure that Johnson was aware of their relationship. Wiesner told me that at the key moment in the meeting, Johnson had pulled out the second Wiesner memo giving the ten reasons for *not* funding MURA, and read it to the group. Johnson did not say that who had written it, or that it was one of two memos. However, after he finished reading it, he placed it face up on the end of his desk, in a way that when Humphrey got up after the meeting, he could see that it was a memo from Wiesner.

In *Fermilab*, Hoddeson, Kolb, and Westfall describe what happened the next day [Hoddeson, 2008]:

"[Elvis] Starr [President of the University of Indiana] wrote an anguished letter asking Humphrey to urge Johnson to reconsider the MURA proposal. The letter complained that the statement read by Johnson 'omitted some highly relevant matters, and its prejudice showed on its face.... "I don't know who wrote it, but it sounded exactly as if it might have been prepared by someone from New England or the West Coast'..." I believe that, the fact that Starr wrote "New England" and not "the Northeast" (after all, the largest competing accelerator was in New York), and put it first before California, MURA's real competitor, indicates that he knew that Wiesner was the author. He did not want to say that someone had sneaked a peak at the memo. (Incidentally, Wiesner had grown up in the Detroit area and was a loyal graduate of the University of Michigan.)

THE POSITIVE IMPACT OF THE MURA DECISION

Although the decision by President Johnson ended the development of the MURA accelerator, it had important consequences for the future development of high-energy physics. The Ramsey Panel Report was credible to Gordon and Johnson, particularly because the Panel was willing to make priority judgments. In explaining their opposition, they both stated as a major factor that there was a higher priority accelerator coming along. Thus the discussions we had in the EOB about the Berkeley proposal over the next four years, no longer dealt with whether it was important to build it. Instead of "Should we build it?", the discussions centered on "How should we build it?", "Where should we build it?", and "How can we do it at less cost?" The MURA turn down thus ensured the approval in the Executive office of the Fermilab proposal.

The MURA turndown also had an impact on the location and governance of the 200 BeV accelerator. It had always been understood that the group that proposed building an accelerator — if their proposal was accepted — would be the ones to build it and use it at a place of their choosing. The MURA turndown indicated that the 200 BeV accelerator would probably be the last large accelerator build for a long while. Therefore, there was a strong view that all high-energy physicists should have an equal chance of running experiments on it. LBL had a reputation of doing almost all its own experiments, with the decision about what should be done made solely by the Director of the Laboratory. Further the new Stanford Linear Accelerator Center used the same management system. The AEC, with OST support, established a national consortium of Universities to oversee the use of the accelerator, rather than leaving it to the University of California.

Again in part, because of the turmoil over the MURA turn down, the AEC decided to have a competition for the selection of the site of the new accelerator, rather than accepting the possible California sites proposed by LBL. A Site Selection Committee was established by the National Academy of Science. It received 126 proposals recommending 200 sites in forty-six states. At the visits that the committee made to the strongest proposals, the local Senators and Congress members who spoke "Took the Pledge" — as one member of the Committee put it.

They would say, in effect: "I strongly support the construction of this accelerator. It should be built here, but if it is not, I will still support it, because of its importance." This process enhanced Congressional support.

Finally, the MURA work over ten years showed that there were a lot of good accelerator scientists in the Midwest. I am certain that this Midwest experience had an impact on Glenn Seaborg, when he and the AEC chose the Weston, Illinois location.

NAME INDEX

Adler, F.   32, 33
Arnold, V. I.   35, 42, 46, 78, 236
Austin, B.   222, 232

Barber, W. C.   66
Barton, M. Q.   69, 75
Binford, T. O.   235
Blewett, J. P.   12, 15, 106, 113, 117, 126, 210, 235
Blewett, M. H.   74, 77, 78, 101, 210, 235
Blosser, H. G.   151, 152, 242
Bohr, N.   86
Budker, G. I.   66, 77, 237
Budker, G. J.   87

Chandrasekhar, S.   86
Chirikov, B. V.   38, 46, 78, 80
Christian, R. S.   19, 28, 34, 35, 126, 217, 218, 232, 242
Christofilos, N. C.   12, 20, 211, 236
Cole, F. T.   1, 5, 15, 16, 18, 27, 31, 46, 62, 74–76, 99, 104, 113, 117, 142, 156, 157, 184, 188–191, 194, 195, 197, 204–206, 208, 209, 212, 213, 215, 228, 232, 242,
Cork, B.   131, 132
Courant, E. D.   11, 15, 16, 32, 33, 74, 79, 101, 154, 156, 184, 236
Courant, S.   16
Craddock, M.   142
Crane, H. R.   10–12, 15–17, 21, 26, 27, 48, 50, 51, 83, 101–103, 108, 120, 125,

133, 142, 149, 156, 184, 192, 228, 231, 242
Curtis, C. D.   72, 79, 121, 123, 125, 128, 214, 222, 232, 233, 235, 242

de Raad, B.   76
del Castillo, G. B.   130, 135, 222, 232

Edgecock, R.   142
Edwards, T. W.   219–222, 235

Fasolo, J. A.   232
Ferger, F. A.   66
Francis, N.   15, 16, 32
Frisch, O.   32, 33

Galonsky, A.   113, 119, 120, 221, 232
Geisler, A.   140
Greenbaum, L.   1
Greenberg, D. S.   1
Guignard, G.   46

Haxby, R. O.   17, 20, 27, 31, 32, 46, 48, 53, 74, 75, 83, 86, 114, 124, 154, 186, 190, 197, 204, 209, 213, 215, 232, 242
Hilden, R. H.   125, 222, 233
Hofstader, R.   16

Johnsen, K.   63, 66, 68
Johnston, L.   15, 16, 32

SUBJECT INDEX